Science, Enlightenment
and Revolution

Science, Enlightenment and Revolution brings together thirteen papers by renowned historian Dorinda Outram. Published between 1976 and 2019 and scattered in a variety of journals and collected volumes, these articles are published together here for the first time.

During her distinguished career, Outram has made significant contributions to the history of science, to the history and historiography of the Enlightenment, to gender history, to the history of geographical exploration, and to the historical uses of language. This volume also includes other writings by Outram, comprising an unpublished introduction in the form of an intellectual autobiography. Placing this together with her collected academic papers offers readers an overview of her development as an historian and a writer.

This book is important reading for scholars and students of early modern Europe, as well as those interested in the Enlightenment, the French Revolution and gender studies.

Dorinda Outram is Franklin I. and Gladys W. Clark Chair of History Emerita at the University of Rochester, USA. She has taught in the UK, USA, Canada, Australia and Ireland and has held visiting positions in Paris, Göttingen and Berlin. She is the author of many works, including most recently *Four Fools in the Age of Reason: Laughter, Cruelty and Power in Early Modern Germany* (2019) and a fourth edition of *The Enlightenment* (2019).

Science, Enlightenment and Revolution

Selected Papers, 1976–2019

Dorinda Outram

VARIORUM COLLECTED STUDIES

Routledge
Taylor & Francis Group
LONDON AND NEW YORK

First published 2022
by Routledge
2 Park Square, Milton Park, Abingdon, Oxon OX14 4RN

and by Routledge
605 Third Avenue, New York, NY 10158

Routledge is an imprint of the Taylor & Francis Group, an informa business

British Library Cataloguing-in-Publication Data
A catalogue record for this book is available from the British Library

Library of Congress Cataloging-in-Publication Data
Names: Outram, Dorinda, author.
Title: Science, enlightenment and revolution : selected papers, 1976–2019 /
    Dorinda Outram.
Description: Abingdon, Oxon ; New York, NY : Routledge, 2022. |
    Series: Variorum collected studies | Includes bibliographical references
    and index.
Identifiers: LCCN 2021013882 (print) | LCCN 2021013883 (ebook) |
    ISBN 9780367481193 (hardback) | ISBN 9781032064543 (paperback) |
    ISBN 9781003038085 (ebook)
Subjects: LCSH: Science—France—History. | Enlightenment—France. |
    France—Intellectual life—18th century. | France—Intellectual
    life—19th century. | France—Civilization—1789–1830.
Classification: LCC Q127.F8 O98 2022 (print) | LCC Q127.F8 (ebook) |
    DDC 509.44—dc23
LC record available at https://lccn.loc.gov/2021013882
LC ebook record available at https://lccn.loc.gov/2021013883

ISBN: 978-0-367-48119-3 (hbk)
ISBN: 978-1-032-06454-3 (pbk)
ISBN: 978-1-003-03808-5 (ebk)

DOI: 10.4324/9781003038085

Typeset in Times New Roman
by Apex CoVantage, LLC

VARIORUM COLLECTED STUDIES SERIES CS1101

CONTENTS

ACKNOWLEDGEMENTS

Thanks to Houghton Mifflin Harcourt Publishing Company for permission to quote from the following:

"The Onion" and "Writing a Resume" from *View with a Grain of Sand: Selected Poems* by Wisława Szymborksa, translated from the Polish by Stanislaw Baranczak and Clare Cavanagh. Copyright © 1995 by Houghton Mifflin Harcourt Publishing Company. Copyright © 1976 Czytelnik, Warszawa. Reprinted by permission of Houghton Mifflin Harcourt Publishing Company. All rights reserved.

INTRODUCTION

Reflections on an intellectual life

> Thus is man that true and great Amphibian, whose nature is dis-
> posed to live not only like other creatures in divers elements, but in
> divided and distinguished worlds.
>
> Sir Thomas Browne, *Religio Medici* (1643) I, 34

I have lived for most of my life on edges and margins, in divided and distinguished worlds, and whatever contribution I have been able to make to historical
understanding has come from those liminal experiences. The academic papers and
other pieces collected here reflect understanding gained in many different places,
Australia and Berlin being the most important; they also reflect understanding
gained by working on the margins of many different historical fields.

Beginnings, however, were standard for my generation of historians. I received
my initial professional training at the University of Cambridge, then in a dominant
position in the discipline, in the late 1960s and early 1970s. Neglecting the struc-
turalist and post-structuralist debates raging over the Channel, it was a training
largely in political and administrative history, which emphasized high standards
of accuracy, respect for facts, and tenacity in seeking them out in the archive. It
also privileged clarity and accessibility in exposition. This is not a heritage with
which I felt obliged to quarrel. The experience of being an undergraduate and
graduate student at a university where the gender ratio was 1:10, with my college
lying two miles out beyond the city centre, also gave me new understandings of
marginality.

However, the great virtues of this early training were also linked to factors
which made it increasingly unsatisfying. Political history of the kind we were
taught seemed to privilege a view of human experience as being totally contained
within power relations, transactions of superiority and subordination, a set of
games in which the powerful manipulated the subjected, and the subjected in their
turn invented manoeuvres to outwit the powerful. Besides the clear and unhelpful
cynicism of these presuppositions, this way of thinking was in the end profoundly
unhistorical and conservative because it offered no way to explain why human
behaviour should ever change. In writing my PhD thesis on the collaborator class

DOI: 10.4324/9781003038085-1

which supported the Napoleonic Empire, I had created such a cynical world. It ended by disgusting and repelling me and moving me into a flight from the profession. I had never been required to ask questions about the public role of the profession.

I was saved, however, from retraining as nurse, plumber, or electrician, professions which I came to feel made a more useful impact, by friends who mentioned to me the existence of a new subfield, that of the history of science. I turned to it with relief in the late 1970s. History of science in England was only newly institutionalized as a discipline, and was not even part of the History Tripos at Cambridge. It was also experiencing a conceptual explosion. Social constructivism and Marxism battled together. Later, theories of representation assumed importance, and questions stemming from feminism began to be asked. Possibly the most heavily theorized of the subfields of history, began to be, in my view, the intellectual area which most rigorously and imaginatively approached basic problems in our relations with past time. To think about man's relation to nature as having a history made the game more complex than that of political history. I brought to the field, at that point dominated by practitioners trained in the natural sciences, the contribution of a historian trained to see change in large political terms. What I gained from it was immeasurable in terms of a demand for rigorous conceptualization, a self-conscious awareness of how historical scholarship may intervene at the interface of fact and value, and an expansion of the questions deemed permissible to ask of human experience. This in turn revived and deepened my flagging appreciation of political history and permitted me to understand it anew in a more positive way.

To work in the history of science also, however, put me in a no-man's land. Historians trained as I was at this time rarely included the history of science in their accounts of society or politics. Conversely, historians of science, usually trained in the sciences, assumed that they could work without the frameworks of general history. "How can you do work in the history of science when you don't know any science?" one eminent specialist asked me. "How can you work in the history of science when you don't know any history?" should have been my response.

It was very clear to me that I could either remain on the margins of both disciplines, or I could inventively use that very marginality to demonstrate how history and the history of science could each exert leverage on the other's agendas, methodologies and master narratives. It was this agenda that lay behind my first significant contribution to the field, my study of Georges Cuvier, which appeared in 1984. This was the history of a marginal man who rose to be the most important natural scientist and certainly one of the most important patrons in France in the period of the French Revolution and Napoleonic Empire. Through his career I explored the workings of patronage systems and institutions in the overlapping fields of science and politics. Power and its brokerage were my main themes. This broke with long-standing assumptions in the history of science, which saw science as progressive and unpolitical. The book shook up the field. The first three papers collected here were offshoots of that writing. All of them demonstrated the important contribution that could be made by placing the sciences of the revolutionary era against the politics and society of that time, by seeing science as one occupation amongst many and, far from being stably progressive, as tossed in the storms of the 1790s. I also consistently argued that scientific institutions, hitherto the focus of the history of science, could only be understood as the creatures of the patronage networks which pervaded them. This made institutions in general, and certainly those of science, unlike the modern institutions to which they had been uncritically assimilated. This was an insight I carried forward into my study of the Jardin des Plantes, "New Spaces in Natural History", also reproduced here. It was new to see institutions as complex spaces where public and private, the professional and the patronage networks interacted. Out of that interaction grew science. One of the major questions I tackled was the relationship between patronage and institution on the one hand and work in science on the other. How did they interact and shape each other? Would we have had Cuvierian science or Lamarkian science without the Paris Jardin des Plantes? Much influenced by the work of the Late Antique historian Peter Brown on the holy men of early Christianity, I also asked questions about the nature of the scientific elite, the new holy men of the modern world, and about the way language held that elite together by shaping a distinctive self-image for itself. These are some of the themes in my articles on the French scientific elites, and they culminate in "Before Objectivity", where I looked at organized patronage as the affair not just of male patrons but of their wives, the keepers of their salons and the recruiters of their denizens.

My substantive work at the boundaries of history and history of science has thus looked upon science as one among – and perhaps the most distinctive one (apart from history writing itself) – of the cultural and social forms characteristic of the European experience. I view science as a culture and a set of practices, which has not only shaped man's relationship to nature but has also created much of the vocabulary and procedures of governments, bureaucracies and legal systems. Values such as 'objectivity', 'rationality' or 'facticity' and procedures concerning truth and evidence lie at the junctures of science, law and history. Political events since then have showed us how important these values are. Lastly, science, I feel, has also been one of the most important producers of novelty. It is produced, that is, not only in relation to past and present but also in relation to the future.

As the sole support of a young family, I accepted a position in the Republic of Ireland in 1984. This experience was brightened by the friendship of the medievalist Jennifer O'Reilly, whose early death is lamented, and by that of the Irish historian Tom Dunne. This was a strange time intellectually and personally, 'locust years', a time when I tried to find out what the price was for acceptability in Ireland and at the same time struggled to find a way to bring previous historical agendas to bear on the unfamiliar material which lay around me. Irish history at that time struggled to emerge from its nationalist past. But simply inverting nationalist agendas did little to open up the profession to the wider world. One way or the other, it was always the British and United States. How was I to bring my knowledge of the history of science and the history of Europe to bear on this situation, especially as it rapidly became clear to me that the history of science, like science itself, had very little presence in Irish culture? The two essays reprinted here show me trying to place a purchase on the Irish situation and suggesting how another crossroads for change could be understood.

It was, however, the case that if Irish culture largely ignored science, it certainly did not ignore literature, and my own lifelong interest in marrying language, literature and history was allowed free rein. My lengthy review of the Irish novelist John Banville's *Newton Letter*, is reproduced here, as an example. I would now emphasize, as I failed to do when this paper was published, the doubleness of the text. Its setting deludes the reader into thinking this is yet another Irish Big House novel. But the plot and even the names of the characters – Edward, Charlotte and Ottilie – give us a heavy nudge towards Goethe's 1809 novel *Elective Affinities (Wahlverwandschaften)*, also set in a big house on a landed estate, whose owners, Charlotte and Eduard, are carrying through all the latest improvements and renovations. Their mature and sober love is deeply wounded by Eduard's affection for the far younger Ottilie. Chaos and death ensue. All the moral struggles of the three parties are powerless against the pull of their 'elective affinities', and, like chemical elements, they combine in accordance with the laws of affinity, not morality. Reading Banville against Goethe, the *Newton Letter* becomes ironic. The Big House is not so uniquely Irish, and science and technology are both agents of improvement and the supplier of a profound chemical metaphor for the helplessness of human behaviour. Newton, about whom the narrator is struggling to write a book, is powerless to provide a framework of rationality which could save the characters from their fates, just as he in reality succumbed to madness.

This is the background of my papers on Irish science and literature. But an invitation to teach history and history of science in Queensland in1989 swung me abruptly from the closed spaces of the Irish world to the boundlessness of Australia. It was there, in another place far removed from metropolitan centres, that a radical change took place in my outlook on history and on the world in general. Stepping outside the aeroplane that first evening of arrival, breathing the soft, clear, scented air of the Pacific, and seeing the stars above me in unfamiliar constellations, I immediately grasped that I had come to a place like no other in my experience. I was astonished by it. The vast spaces of the island continent were exhilarating to experience, and the freshness and depth of historical questioning which I encountered, all led me to the conclusion that I should pack up and send 'home', the notes and drafts I had brought with me (now lost) for a book on bureaucracy and elites in the Napoleonic Empire. This extension of my PhD thesis, rooted in my work on French science, was abandoned without pain under the enormous impact of the strange and new world I and my son had arrived in. For the first time in my life, I allowed myself to be swept up in a place. I realized I knew nothing of the history of the Pacific and set myself to learn it, greedily reading archive and primary sources of exploration in particular in the extraordinary Queensland State Library, a bright glass tower overlooking a broad river. But above all I relished the new sky, the enormous spaces and the resolute attempts of

Australian historians to come to grips with their country's often violent history. Out of all this struggle was being born a culture that had ceased to be colonial and white and that recognized itself as bound evermore to land and space. Something neither colonial nor Aboriginal was in the process of emergence, a new cultural whole, into which history and history writing could decisively intervene. Historians like Inga Clendinnen spoke to both public and profession, as did Greg Dening, the great historian of the Pacific islands.

All of them, and especially Dening, were profoundly linked to place. Dening writes that he personally made landfall on every island he wrote about. His classic work on the history of the Marquesas Islands, *Islands and Beaches*, evokes land and people, myth and seascape in a single unified whole. Because he evokes place so sharply, his account also carried the tragic history of the islands after their 'discovery' into a moral realm explored by all his subsequent work. This linked up with my pre-existing admiration for historians like the nineteenth-century Harvard Professor, Francis Parkman, who claimed to have climbed the Heights of Abraham whilst writing on the French and Indian Wars (the period of what Europeans call the Seven Years' War [1756–1763]). Dening, by vocation a priest, then an anthropologist, and only thirdly a professional historian at the University of Melbourne, brought all the expertise from these three different vocations to bear on his account of the islands. This is the only historical work I have ever read all the way through, without stopping, and begun again at the beginning. I was also fascinated, sitting in my sunlight office in Australia, by his account of the Bligh mutiny and the ships Bligh sailed in, the *Bounty* and, after the mutiny, the Bounty's longboat. By minutely measuring off spaces, at the same time discussing the ritual meanings to White sailors and to Islanders of places on the ship, Dening restored my own feeling for space to me. It was something new for me to learn, as up until then, apart from a single chapter on revolutionary Paris in my 1984 book on Georges Cuvier, place had somehow been taken for granted in my work, as it was in my life.

As an example of passages that moved me, I cite from *Islands and Beaches* the description of the island of Taiohae, one of the Marquesas:

> There are few visitors to Taiohae who have not remarked on its overwhelming beauty. The bay is entered by a narrow pass in the cliffs marked by two rocks rising sheer out of the sea. They were called Mataou and Motou-Nui by Enata, the Two Sentinals [*sic*] by seamen. Behind them the bay itself is a huge ampitheatre with bare rock face stretching above the line of trees. The lower slopes are green, more uniformly now than they used to be, because introduced acacia has run riot over the original varied vegetation. The mountain ridges facing the entrance rise almost vertically to three thousand feet, and slope away in jagged spiny spurs to encircle the bay. Their towering nearness distorts the perspective of the bay, closes it in, shortens the distances. A small hill jutting into the waters of the bay divides the eastern section with its white sand beaches

and its small plain from the more extensive western section with its peb-
ble shore. This western section is cut by two small streams and in two
places along the shore small rocky points nudge into the bay. Behind,
several low spurs prop the mountain and divide the western area into
four small valleys.[1]

Australian historians have to come to terms with the land, given that their land is
one of extreme and violent beauty. Because they live on an island, they come to
terms with islands as well. Let us hear Dening again. In a passage which I must
have read a hundred times, he remarks that the Pacific:

> is a total island world. In that immense sea there are twenty-five thousand
> islands. Every islander has had to cross a beach to construct a new soci-
> ety. Across those beaches every intrusive artefact, material and cultural,
> has had to pass. Every living thing on an island has been a traveler. Every
> species of tree plant and animal on an island has crossed the beach. In
> crossing the beach every voyager has brought something old and made
> something new. The old is written in the forms and habits and needs each
> newcomer brings. The new is the changed world, the adjusted balance
> every coming makes. On islands, each new intruder finds a freedom it
> never had in its old environment. On arrival it develops, fills unfilled
> niches, plays a thousand variations on the themes of its own form.[2]

These are the words of a historian whose society grew out of encounter on a beach
at Sydney Cove in 1789.

Australian historians, above all, were the first English-speaking historians I
had encountered who were explicitly struggling to come to terms with historical trauma. The collapse of Aboriginal culture under White pressure from 1789
onwards and the deadly hostility between the races, resulting in the decimation
of Aboriginal population, were well to the fore in every piece of historical writing I encountered there. History had a public function to remind its readership of
the historical responsibility which had accompanied the establishment of White
dominance and to make that responsibility an integral part of civic culture.

In this collection, pieces on Berlin, which I visited for the first time in 1995
as a result of a lucky invitation to the Max-Planck-Institut for the History of
Science, chronicle an equivalent awakening in an utterly different but almost
equally powerful place. Both Germany and Australia are coming to terms with
historical atrocities – atrocities, in the German case, for which Berlin was the
central planning post. Berlin is a city whose beauty is ordered and at the same
time can be sudden, hidden by high walls and, round the corners of seemingly
boring streets and in the city graveyards, heart-stopping. It led me into the history
of the heart.

Here too in Berlin as in Australia, historians struggle with past atrocities.
There is no space here to recount the enormous historiography which has grown

up around this mission. Suffice it to say that physical references to the Holocaust surround one at every turn. Even the very paving stones beneath one's feet may be engraved with the personal histories of those who lived there before deportation, death or, in rare cases, escape. These *Stolpersteine*, or 'stumbling stones', do indeed upset the regular footsteps of passers-by, the stones of the city, otherwise so often concerned with the Imperial monumental, marking the histories of otherwise unknown individuals. History in Germany is longer and deeper and denser than it is in Australia. Yet the profession in both countries struggles to shape civic consciousness and public responsibility, standing as a bulwark against repetition.

Two papers collected here relate to this sudden deepening of historical perception. My paper whose figurehead is the Greek hero Perseus is a distillation of my thinking about movement, travel, perception and thinking. We had flown for 27 hours to arrive from London into the 'new world' of the Pacific, and the result had been a sudden opening of perception. It was all very far from Cambridge, and I wished to reflect on movement and place change as catalysts of thought. Hence the packing of all my notes on European bureaucracies into a plastic bag which was sent home in the mail and my turning to something else entirely. The other paper, "New Spaces in Natural History", represents another leap, this time into realizing that space was not just an outward phenomenon but that we all also live in inner spaces. Encountering the British artist, essayist, diarist and psychiatrist Marion Milner was decisive here. As she writes:

> I have become deeply aware of the double aspect of space, this outer one that surrounds me, this room in which I am writing, lit by the flickering wood fire and the lamplight, and this inner one which is the space my body takes up, and is not lit at all, it's dark and incommunicable in words, indescribable – but not empty, it's warm and rich, full of an odd sort of joy, though a profane kind, almost.[3]

Milner's words warned me that there were whole areas of human experience which yet lacked their historian. This was a perception which I had tried, inadequately and haltingly, to come to grips with in my *The Body and the French Revolution* of 1989: the perception that 'body history' almost always lacked this vital dimension of inner space. How to recover it is one of the mysteries of historical writing. Writing about the Stoic self-perception of the men of the Revolution took me only some of the way there. And recovering it was important in the context of this book because I was making the argument, however haltingly, that the authority of the body was vital to its public role and that the diminished authority of the body in our own age had helped to usher in the ages of Fascism and Nazism and might account for many of the difficulties of democracy. The Stoicism which had lifted many to heroic status in the age of the French Revolution now has little purchase in the public realm.

7

Milner's words also secured my own growing feeling that my inner and outer worlds could and should come together in my own writing and thinking. This search after the internalization of knowledge was to take me on many journeys into what Novalis calls the "inner Africa". Sadly, I also perceived that to believe this was to put myself once more on the outside of the profession, and in a far more fundamental way than when I turned from general history to the history of science. This remained something of a private perception as increasingly over these years, from 1995 onwards, my time was taken up with the writing of the very expository prose that Cambridge had taught me so well.

My textbook on the Enlightenment, originally written on the kitchen table in Ireland, now in its fourth edition, kept me anchored. It also naturally allowed me to think far more closely about the nature of the Enlightenment, its structure and importance. As I wrote and published and (at the request of Cambridge University Press) wrote and published my textbook several times more, each time with revisions and additions, my understanding of the Enlightenment changed. Once interested in the Enlightenment primarily as the forerunner of the French Revolution, I began to understand it as an intellectual world irretrievably marked by paradox and social fear. Its vaunted critique of spiritual tyranny and legal injustice, slavery, luxury, and religious intolerance, to name only a few, stopped short at implementation in almost every instance. The proclamation of the dignity of man was compatible with fear of the end of slavery; Kant in a famous essay queried the consequences for social order if Enlightenment were taken to its logical conclusion of unlimited critique.

When revolution broke out in France, it was blamed on the intellectuals such as Voltaire and Rousseau, not only in the 1790s but right through the nineteenth and early twentieth centuries. As I wrote on, I also became aware of the extent to which the Enlightenment had become fodder not only for historians but also for social theorists and philosophers, from Theodor Adorno and Max Horkheimer to Richard Rorty. A recent essay from 2019 collected here, 'Enlightenment Struggles', deals with this in more detail. Suffice it to say that this afterlife of the Enlightenment, which it enjoys to a greater extent than any other historical period, has made it the plaything of both liberals and conservatives. The recent spate of books reinterpreting the Enlightenment as the origin of the current crisis of liberalism, is typified by Anthony Pagden's *The Enlightenment and Why We Still Need It*. Pagden, whose work I have also criticized in the pages of *Four Fools in the Age of Reason*, sees Enlightenment political theory as the origins of new world based on sympathy among cosmopolitan elites. Blind to the Enlightenment's difficulties in finding ways to include the classic outsiders, unchanged since the Greek polis – women, slaves, foreigners and dark-skinned peoples – he considers it a quarry for the readers he refers to as 'us' and 'we', without ever defining their characteristics, for his new world of cosmopolitan sympathy. But the difficulties are as important as the Enlightenment's aspirations. Without taking them into consideration, Enlightenment dissolves into paradox.

I also objected more formally to these and similar efforts to make the Enlightenment our contemporary in the paper reproduced here with that title. The paper seemed to me to denote a possessive attitude to history in general. This could only result in a lessening of the perception of the distance between us and our past and a lessening of what has been called the 'strangeness' of the past. The past is not always usable fodder for the present. It may, much more, remind us of the very 'strangeness' and plasticity of human culture and human beings, plasticity which current neurological research on the interface between brain, nervous system and meditation seems only to emphasize.

In interesting myself as much in the interpretation of the Enlightenment as in its positive history, I was picking up again the view that my early immersion in the history of science had taught me: that history was valuable, amongst other things, because of its capacity to stand at the intersection of fact and value. Rather than bellow my interpretation of the Enlightenment (for a textbook is not and should not be written with the hauteur of personal interpretation), I chose to embody it in my most recent book, *Four Fools in the Age of Reason: Laughter, Cruelty and Power in Early Modern Germany*.[4] This book was a long time in the making. For more than ten years, I dragged together materials on court fools throughout Europe in the early modern period. These materials remained an inchoate mass for long time. Pulled aside by other projects such as the successive editions of the Enlightenment textbook and *Enlightenment Panorama*, I hesitated to make a definitive beginning. At length, I identified a peculiar incident involving a fool, Solomon Morgenstern, employed in 1737 by the King of Prussia, Frederick William I, to humiliate the recalcitrant professors of the University of Frankfurt-an-der-Oder. In the course of so doing, the fool formally debated with the assembled professors on the meaning of foolishness and reason and published his own theses in a much reprinted pamphlet. That pamphlet made me realize that in talking of fools I was tapping into a whole world of foolishness, composed of riddles, irony, paradox and folklore, which was well recognized by all who were exposed to it. In thinking about the fool's debate with the professors under the king's direct auspices, I also realized that foolishness could be part of the artillery of power used by Frederick William to hustle and shame his recalcitrant elites into obedience. But the real revelation came in conversation with a German friend, the historian Martin Gierl, who simply said, "Dorinda, this is not an article, it is a book". I went ahead and wrote the book, concentrating on the eighteenth-century German states, each one of them, far from the sleepy territories of older historiographical legend, living laboratories trying experiments in the making and shaping of power. I pulled together the lives of four very different fools in four very different regions of Germany and found them interacting with monarchs and rulers in these laboratories of power and culture.

But the fools died. There is no equivalent role in modern life, for the comics and commentators who play to mass television audiences do not face the tightrope walked every day by the fools between advice, laughter, and lèse-majesté.

I devoted some time to trying to find out why the fools died out, and with them their riddling, paradoxical, 'folkloric' way of talking and thinking. Now, there seem to be few who speak truth to power, and few who use paradoxes. This is important because paradoxes disturb perception. They ask us to hold, simultaneously present to ourselves, two opposing things. There is a chance that by doing so we understand more of the deep dependencies between apparently incompatible elements. In particular, I began to understand that the wise men and intellectuals of the Enlightenment, the philosophes and *Aufklärer*, were the paradoxical opposites of the fools. Calling for reason, toleration and progress, they seemingly could not be more different. Yet the central Enlightenment paradox is that both reason and foolishness spoke truth to power. The fools died out when the courts changed with the French Revolutionary and Napoleonic wars. They are a loss for us all. They also died out due to the rise of a middle-class culture based on print rather than performance, on restraint rather than involvement, on decorum rather than truth. This rise has been documented by historians of the 'Moral Weeklies' much read by the middle class and by historians of the rise of moral audiences to complement the bourgeois moral drama which replaced both buffoonery and high tragedy on the German stage by the end of the century.

It is easy to see that the process of writing a book about court fools would radically have changed my ideas about the nature of the Enlightenment. In tandem with work by Erik Midelfort, I began to see the Enlightenment as a top-dressing of important ideas, about justice, progress and toleration, to name only a few, many of which percolated through literate society, as I tried to make clear in *Panorama of the Enlightenment*.[5] Yet surrounding and co-mingling with these ideas were whole worlds of folk story and ritual, nonsense tales and proverbs, joke books at the expense of the learned, exorcists and dream interpreters, popular plays represented at fairs by travelling troupes, broadside ballads and chapbooks, all of which had nothing to do with the so-called high Enlightenment of Voltaire and Diderot. I also saw Enlightenment as profoundly crevassed by gaps between intention and action, as dedicated to ideas but simultaneously seeing those ideas as dangerous to the public tranquillity. Unlimited critique could destroy the social and political order. Gender and racial inequality also undermined the Enlightenment idea that human beings were distinguished by the use of reason. As Mary Wollstonecraft pointed out, an Enlightenment that denied to other races and women the possession of their reason was an Enlightenment in name only. I end my textbook's most recent edition with a too-short evocation of the ordinary men and women who struggled to make "Enlightenment models shape their self-consciousness, who thought through Enlightenment debates, and found satisfaction in thinking of themselves as enlightened people".[6]

I was struck in my reading with the way Enlightenment people often described their reading experiences as 'digesting' a book. In other words, they internalized it, by reading such bestsellers as Goethe's *Sorrows of Young Werther* many times and by furiously identifying with hero and heroine alike. I came across young people who laboured to form their identities on the basis of John Locke's philosophy. In 1741, for example, Eliza Lucas, (later Eliza Pinckney) a bright, self-educated and independent-minded adolescent, living in the planter society of South Carolina, became concerned at her rapid mood swings. She knew where to find discussion of her problem. "I was forced to consult Mr. Locke over and over, to see wherein personal identity consisted. And if I was the very same self", she wrote, showing that a hard encounter with the English philosopher could lead to the fulfilment of the ancient maxim "Nosce te ipsum" or "Know thyself".[7] This was particularly important in an age resonating with the Scottish philosopher David Hume's remark that "I may venture to affirm of mankind, that they are nothing but a bundle or collection of different perceptions, which succeed each other with an inconceivable rapidity, in a perpetual flux and movement".[8] Enlightenment could involve hard work on the self. That hard work was undertaken by breast-feeding mothers mocked for their adherence to the ideas of Rousseau on family life, or the reading of lonely autodidacts in remote villages, or the self-education of slaves, servants, and young girls too bright for their surroundings.

It may be, I find, that the whole concept of the Enlightenment has simply exploded and that no consensus has emerged to take its place. The term may have become nothing but a catch-all for the very diverse thoughts and practices of the eighteenth century. Yet the Enlightenment was explicitly, to very many people, something that mattered. They wanted to be enlightened. Enlightenment was the focus of energies, needs and desires. Its aspirations such as the achievement of religious toleration, the abolition of judicial torture, the belief in human progress, the critique (however limited) of Church and state, the remaking of the family, all acted as spaces where women and men could validate their status as members of a new elite of thought and word rather than of birth and wealth. The Enlightenment, in spite of all local and national variation, was a project to harmonize the cacophony of the world and to explore the paradoxes of human needs and passions. But to get there was to engage in struggle and process. Ideas did not find instantiation at the moment of their utterance. My ideas about the Enlightenment have thus greatly changed between edition one and edition four of the textbook.

Also in writing about the fools, I came to grips with the uses of unreason, of cruelty and laughter directed at the fools, sometimes manipulated by them and sometimes not, in creating political power for rulers through the consolidation of elites at court. This was a far cry from the rational debates I described in my textbook as characteristic of the eighteenth century. Manipulation through laughter and cruelty both were commonplace in the German courts I described. *Four Fools*, however, had additional objectives. Perhaps too demurely, it took issue implicitly with history writing's own search for rationality, its assumption that the game of calculated advantage was the sole subject matter. This is an assumption made even by the anecdotal historians of the New Historicism, from whom I delicately tried to separate myself in these pages. The book on fools looked at first sight like ideal New Historicist material, loaded with quaint anecdotes and dealing with a past decidedly full of 'strangeness'. But it was not my intention to make the fools or their masters quaint. It was, partly, to explore the power of marginality. Fools were marginal people. They were there to be mocked and manhandled. But they also gave advice to princes, told truth to power. They used princes for their own advantage and were used by them to shape their elites through mockery, at a time when the reshaping of elites was the way in which absolutism stayed in power. The fools worked within a culture of foolishness which went beyond the games of calculated advantage in order to turn the world upside down and to hold apparently incompatible things present to each other through paradox, contradiction, and riddles. They embodied the non-rational, the marginal.

Writing the book on fools involved bringing together many different sorts of material as evidence. These materials ranged from the autobiography of the Tyrolean itinerant fool Peter Prosch, to stage settings for the buffoons who took the culture of the fools to admiring and appreciative theatre audiences. My two major secondary authorities, were both written by people who were marginal in important ways. First comes the history of fools and foolishness published (posthumously) by Flögel in the world-shattering year of 1789, which also saw not only the opening of the French Revolution but the first white Australian settlement on the beach at Sydney Cove. This work of heroic compilation is a mine of information on fools from the Greeks and Romans onwards, with special emphasis on Flögel's own immediate past generation. An elegy on the passing of the fool, it is also a lifetime's work of critical history, sifting and sorting anecdotes and documents on a vast scale, crossing boundaries that we would now hold sacrosanct between classics, modern history and literature from elegies to joke books. Discussion of court fools leads to digressions on statecraft.

Flögel himself combined elements of marginality and authority. A provincial *Aufklärer*, headmaster of the gymnasium where he had himself gone to school, a translator from English, a correspondent of Herder, a published author of several other multivolume books on other aspects of comedy, he combined his literary authority with marginality in place – Liegnitz in Silesia – which cut him off from the intellectual hub of the Berlin Academy of Sciences or the leading universities such as Göttingen. What shines through his writing is both humanity and nostalgia, which see comedy and foolishness as essential to human existence, and the decline of the fool as a loss for the human race.

My second major authority was very different from Flögel. Enid Welsford, whose life just overlapped with my career as a Cambridge undergraduate, originally became famous as a child poet in the years before the First World War. After publishing, in the 1920s, a classic book on the English court masque, she produced the book which influenced *Four Fools*, her well-known 1934 *Social History of the Fool*, which was reprinted and which formed the basis of several stage plays.[9] Like Flögel, she runs over boundaries, combing a brilliant analysis of the Fool in *King Lear*, in one chapter, with analysis of Erasmus's *Praise of Folly* and details of the historical existence of Italian and German fools. A Fellow of Newnham College for most of her life, she revelled in the intellectual ferment brought to the College by Katherine Harrison and others like Nora Kershaw and by her husband Hector Munro, who worked on anthropology, ancient poetry, and folklore, always on the edge of historical study. All of these interests mark the book on fools. Like Flögel's, Enid Welsford's outlook is nostalgic. Writing in the aftermath of the First World War, in which she lost a beloved brother, and on into the age of the dictators, she wrote eloquently on how the fools could have injected a healing laughter into the excesses of fascism:

> It is not quite clear that the change of mental climate which was so fatal to the fool is proving in the long run altogether wholesome to ourselves. [For] . . . who is to laugh at us and remind us of mortal inadequacy? Who is to present our humanists and dictators with the cap and bells? A good many things have been said and done in the last century and in our own day to attract the Evil Eye; perhaps a little more nonsense and self-mockery might have brought us better luck.[10]

A staunch Anglican, her conservative, nostalgic outlook led into a high valuation of the fools as important carriers of a whole range of the symbolism of order and majesty fast vanishing from her own world, yet she was also able to praise their riddles and paradoxes born of their own marginality.

Nostalgia does not interfere with the classic fool's errand of telling truth to power. Indeed, it may be a condition for it. This is one of the reasons why the history of fools is important. In our own day, nostalgia is a forbidden emotion, one that dare not speak its name, seen as equivalent to self-indulgence and intellectual laziness. Yet what else informs the almost obsessive and possessive use by modern historians and philosophers of the eighteenth century as a basis for speculation about the future? Hence the two papers collected here about nostalgia for the Enlightenment, the most recent one from 2019 concerned with Enlightenment historiography and its political implications.

Welsford was a Fellow of her College and, for many decades, University Lecturer in English. Like Flögel's, her institutional basis was solid. But as a woman at Cambridge, she was marginal in a university which refused to grant degrees to women until 1946 and where even after that date the gender ratio was 1:10 in my lifetime. The women's colleges occupied physically marginal spaces, my own being more than two miles outside the city centre. Her mixing of genres, her interest in poetry and language – her last book was on Wordsworth – and their relation to history, struck a responsive chord in me. In many ways, she was an unseen mentor, whose Cambridge life as an unmarried woman scholar I could identify with and whose movement of interests over her lifetime from Anglo-Saxon poetry to Wordsworth I could assimilate to my own constantly changing interests from book to book, place to place. The next section of this Introduction looks more concertedly at the issue of friends and mentors, seen and unseen, and how they have influenced the papers gathered here, and the books written between the papers.

II

Life is like an onion, composed of many different layers, some tasty, some leading to tears. Some years ago I fell in love with a poem about an onion which says this more eloquently than I could. Written by the Polish poet Wisława Szymborska, who won the Nobel Prize for literature in 1996, it goes (in translation) as follows:

> The onion, now that's something else. / Its innards don't exist / Nothing but pure onionhood / Fills this devout onionist. / Oniony on the inside, onionesque it appears. / It follows its own daimonion / Without our human tears. / Our skin is just a cover-up / for the land where none dare go, / an internal inferno, / the anathema of anatomy. / In an onion there's only onion / from its top to its toe, / onionymous monomania, / unanimous omninudity. At piece, of a piece, / internally at rest. / Inside it, there's a smaller one / of undiminished worth. / The second holds a third one, / the third contains a fourth. / A centripetal fugue. / Polyphony compressed. / Nature's rotundest tummy, / its greatest success story, / the onion drapes itself in its / own aureoles of glory. / We hold veins, nerves and fat, / secretions' secret sections. / Not for us such idiotic / onionoid perfections.[11]

My first publication was a poem, not a historical paper. Written under the influence of a peerless English teacher, at around age 17, I have it still. It was published in the school magazine, which has survived all the packing and unpacking, shipping and unshipping, of a vagabond existence. All my life I have played with words. I have used poetry in teaching history, valuing it for its capacity to evoke empathy, provoke thought, encourage delving below the surface of language, and deliver 'strangeness', all at the same time. For many years I opened my graduate class with a poem, often a poem about time, love or exploration. And for many years, I invited students to reflect on Keat's famous idea that for the poet, a great gift is "[n]egative capability, that is when a man is capable of being in uncertainties, mysteries, doubts, without any irritable reaching after fact and reason". Of course, in the end, the historian writes with fact and reason. But the way thither may well resemble far more closely the poet's groping after form. Greg Dening reaches new levels of honesty for the profession when he writes about

> the bleak days, the chance discoveries, the opportunistic sallies, the fumbling questions and the occasional rewarding days of light and excitement when understanding and new knowledge reward the effort it takes to uncover one tiny corner of the universe of truth.[12]

For many years I maintained in Rochester a wall anthology of poems from Graves, Skelton, Marvell, Shakespeare and that great English poet Anon, closest to the life of the people, and guardian of the magical, penetrating and strange. This is why I discuss poetry in this volume, not only because it says things more eloquently than I am able to, but also because of its potential effect on the young. As has recently been remarked, "A more indirect poetic mode of writing can force readers to question their deepest assumptions. . . . Poets tend to be attracted to words, while scholars often prefer terms . . . [but] termini lack the parola's genetic bond between word, world, memory and time. They denote but do not, as it were, arouse".[13]

Perhaps my first unseen mentor was the poet Robert Graves, out of fashion now, but a master of the romantic lyric, as well as of the historical novel, such as *I Claudius*, for which he is better known to a wide audience. Graves, born in 1895, was still alive and writing when I came to consciousness at around the age of 15, and I devoured not only his poems and novels but also his manual (written with Alan Hodges) of prose style, *The Reader over Your Shoulder*. Whatever I have of a crisp, clear muscular style, which takes account of the deep history of the language, I learnt from Graves. His consciousness of the history of English poetry and of the poetic vocation in the British Isles, from the Irish court poets and the Welsh bards onwards into the eighteenth century, which he saw as a period of grovelling decline in poetic independence and the abandonment of muse-driven poetry, fed strongly into my own struggles to find vocation, itself a topic of the early papers presented here.

Graves was my muse, in spite of all his warnings of the danger of a male muse for a woman. As all adolescents do, I wrote poetry, all of which I later destroyed. And as a few scholars do, I found a double muse in the union of Clio for history and Thalia for comedy and idyllic poetry. Under their combined aegis, I have written ever since. This will no doubt be condemned as unprofessional. Some words by Graves will say it better:

> A poem is addressed to the Goddess. She smilingly forgives clumsiness in the young or uneducated – early poems have a nap, or bloom, not found in later poems. And she appreciates the loving care put into a poem by the more experienced; she dislikes slovens. But she insists on truth, and ridicules the idea of using argument or rhetorical charm to overbear her intuition of truth. . . . A poet's integrity, then, consists in his not forming ties that can impair his critical independence, or prevent him from telling the whole truth about anything, or force him to do anything out of character.[14]

Prose, it may be objected, is the proper tool of the historian. But they mingle in Graves's argument that professionally minded English poets ban double-talk, except in satire, and insist that every poem must make prose sense as well as poetic sense on one or more levels.

Graves's insistence on the power and necessity of a relationship with the muse for the poet – and I would add, the historian – is of course, going against the grain of the age. As long ago as the early nineteenth century, the inspired poet William Blake lamented, yet paradoxically, the passing of the muses:

> Whether on Ida's shady brow / or in the chamber of the East, / The chambers of the sun, that now / From antient melody have ceased; / Whether in Heav'n ye wander fair, / Or in the green corners of the earth, / Or the blue regions of the air / Where the melodious winds have birth / Whether on chrystal rocks ye rove, / Beneath the bosom of the sea / Wand'ring in many a coral grove, / Fair Nine, forsaking Poetry! / How have ye left the antient love / That bards of old enjoy'd in you! / The languid strings do scarcely move! / The sound is forced, the notes are few![15]

It should come as no surprise that my second unseen mentor was Helen Waddell, the historian of the Late Antique, almost the inventor for the ordinary reader (long before Peter Brown) of this period, its poetry, and its theology. Helen Waddell was born in 1889 in Japan to a missionary father of Ulster origins. By 1901, both her parents were dead. By 1908, she was an undergraduate at Queens University Belfast, and the mentee of the great literary scholar George Saintsbury (1845–1933). Yet her brilliant intellect, command of language, and mastery of her themes never found reward in an academic post. Indeed, she was denied such a post at her alma mater, Queen's University, purely and explicitly on the grounds of her gender. Undaunted, she supported herself by teaching, reviewing, translating and working at the publisher Constable & Co. in London. She survived ten 'locust years' in Belfast during which her time was taken up in caring for a grudging, querulous alcoholic stepmother. The stepmother's death in 1919 and a fellowship to study in Paris enabled her to leave home. *The Wandering Scholars* became a bestseller in 1927, close to Enid Welsford's own first adult publication. She became a public figure, a friend of prime ministers and statesmen, interviewed and feted on both sides of the Atlantic.

Above all, her work brought a distant and difficult period of history, the so-called Dark Ages, to within reach of the common reader. In doing so, she went against the grain of the age in academic studies, spearheaded by I. A. Richards's 1924 *Principles of Literary Criticism*. A revolt against *belles lettres* in literary criticism, Richards concentrated on text rather than context and was a forerunner of the movement for 'close-reading'. Waddell's objective, conversely, was to situate her material in its historical and moral context. Her first book, *The Wandering Scholars*, was an account of the wandering life and Latin poems and songs composed by medieval 'goliards' or clerical students who wandered from university to university to sit at the feet of famous teachers such as Peter Abelard of the Sorbonne in Paris. They drew on their classical and legal education as they composed love lyrics and satirical parodies. Waddell's source was the famous *Carmina Burana*, a collection of goliard songs discovered at the German monastery of Benediktbeurn in 1803, of which the last was written in about 1225. Working on the fine line between history and literature, she contested the idea then current that goliard poetry in late Latin was of lesser literary value compared to lyrics in emerging vernaculars such as Provençal, English or Tuscan. The wandering scholars fascinated her. She called them "greedy of experience, haunted by beauty, spendthrift and generous, fastidious and gross".[16]

In writing about them, she was undertaking, again against the grain of the history writing of her times, to fill in a gap in the universal and enduring history of Roman civilization, offering an implicit protest against the division of history in her day into distinct national, even tribal parts. This was a historiographical tendency present since the days of German Romanticism. It played all too well into the needs of the age of the dictators for the theories of race separation and hierarchy which underpinned Nazism and Fascism. The great literary historian Ernst Robert Curtius remarked in 1948, "To renounce the Latin tongue and the centrality of the idea of Rome is to play with the fire of barbarism". He went further and explicitly mentioned Helen Waddell:

> [T]o transmit tradition is not to solidify it into an immovable body of doctrine . . . or into a fixed choice of canonical books. . . . [T]he study of literature ought to be conducted in such a way as to give the student joy and to make him marvel at beauties which he did not even suspect. Devotion and enthusiasm are the keys which will open these hidden treasures. I believe that wide stretches of medieval literature are still waiting for the divining rod which will point to sources of beauty and truth. The books of Miss Helen Waddell have performed such services for a great number of readers. Thanks to her, the songs of the wandering scholars have found new, delighted audiences.[17]

The scholars wandered in search of better learning, their lives rather like that of the Enlightenment writer Oliver Goldsmith, "who jigged his way through Europe with a flute and a trick of Latin disputation; the flute got him bite and sup from the country folk; his argumentative tongue and dog Latin three days board and lodging in a monastery, for that was the prize of a victory in debate".[18]

Helen Waddell, whose childhood had been spent in wandering between China, Japan and Ireland and who never knew the security of an established academic post, undoubtedly identified with these wanderers. They were not personae gratae to the medieval church. Church councils from the ninth century onwards condemn the goliards. They abused clerical privilege, which gave relief from every tax and imposition of the secular power, and, above all, freedom from trial in a secular court of law:

> *Stabilitas*, perseverance in the place where a brother had made his profession, was one of the three obligations of the Benedictine vow. And

the rule which was absolute for the monk . . . holds, though with more elasticity, for the clerk. . . . No clerk may leave the diocese without permission and letter of license from the bishop; no bishop may receive him without such a letter. . . . [T]he natural state of the clerical soul is static.[19]

If one wished to abuse an antagonist, one called him a *gyrovagus*, or wanderer.

It was the poetry of these wanderers that Helen Waddell opened up to her readership. She invited them to enter a historically grounded world, which was yet of all times and places. As she wrote, "[T]here are no times or countries or languages in the kingdom of poetry". With that belief, she allowed her readers to be both in time and out of it. As she wrote, "Alcuin and Matthew Arnold had heard the cuckoo's parting cry above the shining water meadows alike of Oxford and Touraine, and each had known it for the dirge of the unreturning springtime of the heart, the sorrow of Persephone's garden".[20] In language of great beauty, Waddell managed to release her readers from the constraints of time while opening to them an unknown medieval world. Most readers, including academic reviewers, shared her essentialist assumptions, and few questioned the continuity of emotional impulses across centuries. She wanted to remove "the wall of glass, impalpable and deadening" that separates the present from the past.[21] Above all, she identified with the wandering life of the scholars and saw them against the enormous backdrop of the Latin past: "The scholar's lyric of the twelfth century seems as new a miracle as the first crocus; but its earth is the leaf-drift of centuries of forgotten scholarship. . . . To the medieval scholar, with no sense of perspective, but with a strong sense of continuity, Virgil and Cicero are but the upper reaches of the river that still flows past his door."[22]

Why was the *Wandering Scholars* so successful? Part of the answer must lie in the clarity, freshness and poetic beauty of Helen Waddell's language, and the novelty to most readers of the period she makes available to them. Her identification with the clerks also brought them closer to home to twentieth-century readers. Some historians have argued that in Waddell's wandering, subversive clerks, the readers saw the rootless generation which had survived the First World War and had come home with a distaste for the authorities who had caused the needless death of millions.

The book turned Helen Waddell into a public figure. Her next major work, was a novel, *Peter Abelard*.[23] It was a work again pulled into being by her intense identification with people from the past: this time, the medieval cleric, philosopher and theologian. Abelard is known now, if at all, to the general reader for his ill-fated love affair with his star pupil Héloïse, which resulted in a baby, his castration by her outraged relatives, and her lifelong retreat into a convent of which she eventually became the abbess. From that retreat, she initiated a spiritual correspondence with her former lover which is one of the most well-known works by a woman of the Middle Ages. In a letter, Helen Waddell wrote "The Abelard chapters are pretty good, I think, but upon my word they nearly kill me, for you live the things while you're writing them".

Earlier, Waddell had experiences in Paris which demonstrated the depth of that identification. She became ill there in 1924 and was admitted to the Institut Pasteur. She wrote later in a passage worth quoting at some length:

> I passed, fully awake and not I think delirious, into some strange state of being. For suddenly I was Héloïse, not as I had ever imagined her, but an old woman, abbess of the Paraclete, with Abelard twenty years dead: and I was sitting in a great chair lecturing to my nuns on his *Introductio ad Theologiam*. It was near the end of the lecture, and I pronounced the benediction, and sat watching them go out, two by two. And one of them, the youngest and prettiest of my nuns for whom I felt some indulgence, glanced at me sidewise as she went out, and I hear her whisper to the older sister beside her, "Elle parle toujours Abelard". It stabbed me. And even when the first hurt of it was past, the realization that what was once a glory in men's minds had become an old woman's wearisome iteration, I began wondering if it were indeed true: if after all these years I were lecturing on this theology for the sake of now and then naming his name. And from that I began to remember that his theology had been condemned as heresy: and – for by this time Abelard had done this work upon me and had brought me to some sense of God – I began to wonder if I had periled the souls in my charge by teaching them heretical doctrine for the sake of gratifying this ancient lust. . . . Then the morning came, and with no sense of transition I was myself, but with full awareness of the other who I had been all the night before: and when the Mother Superior came to see me during the morning, I laughed and said, "Ma Mère, I too was an Abbess all last night".[24]

In 1936 appeared the third work which kept her before a devoted public, *The Desert Fathers*, translations from Latin works of the monks who retreated from their large, overcrowded monasteries to isolated cells in the Egyptian desert in search of greater holiness. Her introduction tells the story of the growth of the holiness of the desert through the life of St. Anthony and other desert saints. They were taken from the *Vitae Patrum*, a vast collection of the lives and sayings of the desert fathers, printed in Antwerp in 1615, which she first encountered in Paris in 1924. Waddell offered her readers only a fragment of the vast corpus and modestly remarks that the business of the book "is only with translation. It is not a study of the desert fathers, their place in the ascetic tradition, or the authenticity of the sources".[25]

She continues, again stressing the closeness of her life with the past:

> I came to the *Vitae Patrum* sixteen years ago, not for its own sake, but in a plan I had of reading for myself, with a mind emptied, what the

ordinary medieval student would have read, to find the kind of fur-
niture his imagination lived among. It held me then, as now, with its
strange timelessness. . . . I have left on one side the spectacular auster-
ities which Gibbon and his successors have made sufficiently familiar;
they are the commonplaces of controversy. It is forgotten that inhu-
manity towards oneself had often its counterpart in an almost divine
humanity towards one's neighbor; . . . The desert has bred fanaticism
and frenzy and fear; but it also bred heroic gentleness. This selection,
fragmentary and arbitrary, is neither comprehensive nor quintessen-
tial; but it represents that part of the Desert teaching most alien and
most sovereign in a world that has fallen to the ancient anarchs of
cruelty and pride.[26]

It is easy to see those who retreated to the remote deserts as mere "barnacles
upon eternity", as making little impact on the struggles of the late Roman Empire
to find and keep a civic consciousness, and as being concerned only for their own
personal sanctity. Waddell continues this thought by admitting: "[O]f the depth
of their spiritual experience they had little to say; but their every action showed
a standard of values that turns the world upside down. It was their humility, their
gentleness, their heart-breaking courtesy, that was the seal of their sanctity to their
contemporaries, far beyond abstinence or miracle or sign. . . . [I]n the world, they
taught us how man makes himself eternal".[27]

This argument about the saints of the desert, who also included such great
scholars as the Bible translator St. Jerome, are arguments too about the pivotal
role of solitude. Waddell argued that "the last delusion of the human heart [is]
that solitude is peace. The truth is that solitude is the creative condition of
genius, religious or secular, and the ultimate sterilizing of it. No human soul
can for long ignore 'the giant agony of the world' and live, except indeed the
mollusc life, a barnacle upon eternity". Even the "Desert itself came to realize
that solitude is a thing to be earned, by passage through the world".[28] These
were also words which apply to the life of scholarship. Before 1934, Waddell,
who never married, lived in a series of apartments, her equivalent to the isolated cells of the Desert. In that year, against the advice of most of her friends,
she purchased an enormous house on Primrose Hill in London, which gave
shelter to a drifting colony of friends and parasites, who impeded her work
and put an end to the solitude necessary to it. It was like the same crowded
monasteries from which the Desert Fathers had fled. Even more fundamentally,
she became locked into a lifelong platonic relationship with Otto Kyllman, the
senior partner at her publisher, Constable and Co. Not free to marry, Kyllman
moved into Waddell's house and stayed there for the rest of his life, giving rise
to what Waddell admitted was a "frustration and torture of self-denial", not
unlike the world of Abélard and Héloïse, about whom she had written only the
previous year.

The Second World War precluded any more publishing by Helen Waddell. She gave up her time to war work and never finished her projected book on the medieval scholar John of Salisbury, nor put together a book of her own poems, many of them published in her lifetime in disparate locations, or of her translations from classical and late Latin. The overstress of London under the Blitz contributed to the gradual failure of her faculties after the end of the War. At the end, she lived in a peaceful fog. Unaware of Kyllman's death in 1958, she died in 1965.

What did Waddell's life and achievements teach me? It validated my own life which I felt took place on the edge between history and literature. It validated achievement coming from the margins and from a wandering life. It validated my own demands for solitude, a demand traditionally less allowed to a woman, as is demonstrated by the woman writers such as Jane Austen who never had "a room of their own". It also validated the intense identification with the past as a precondition of writing about it. Identifying with the past also implies a recognition of past time as being past and yet as also being capable to some extent of being brought to survival and leaving traces on the present. This is precisely what separates me from New Historicism. Waddell's allusive style, constantly marrying present and past examples, Alcuin and Matthew Arnold, perturb the surface of her analysis. But they also hold up present and past to each other in a way which grows our empathy.

The last word should lie with Waddell's biographer, Monica Blackett, who emphasizes not only her scholarly brilliance but also her compassion, humour and understanding of simple things and people.: "Her humility took the danger out of flattery; her sense of humour was as fundamental as her instinct to sublimate platonic love, and her own virginity which she retained throughout her life. . . . Pre-eminently she desired to give the whole of herself; all her basic emotional intuitions, all her sublimations, and all her acquired knowledge".[29]

There are many obvious parallels between the lives of Helen Waddell and Enid Welsford. The most obvious difference is that Helen Waddell was a scholarly *gyrovagus*, and Enid Welsford was not. Yet both wrote on that margin where history and literature mingle. Both loved language. Both used language to bring their scholarly knowledge to a broader public, assuming, rightly, that even the most difficult and distant periods and avocations (desert saints, or even fools) could be brought before the broad public with élan. Both were nostalgic writers. Both believed that nostalgia is powerful in its critique of the present.

In conclusion: The taking into one's life of the living and the dead as models and muses demands an intense involvement with individuals. It is not precisely a search for authority, for something higher than oneself, for what occurs is an absorption, a meditation, an identification, not a servitude. Individual authorship, after all, is a polite fiction. In reality, we are palimpsests, parchments where layers of previous writings can still be read and which illuminate the meaning of the neatly penned illusion of authority on the surface. This involvement with other individuals is part of the way in which an authoritative person is built up. 'Being oneself" is a deadly phrase. One has no choice but to be oneself, but that self is an incorporation of many others. If the King has two bodies, we have many. How we live with them makes our own authority.

III

In teaching beginning graduate students, I often used to ask them to contemplate another poem by Szymborska, called "Writing a Resumé", about the process and the price of academic self-representation. Here it is:

> What needs to be done? / Fill out the application and enclose the resumé. / Regardless of the length of life, / a resume is best kept short. / Concise, well-chosen facts are de rigeur. / Landscapes are replaced by addresses, / shaky memories give way to unshakable dates. Of your loves, mention only the marriage; / of all your children, only those who were born. / Who knows you matters more than whom you know. / Trips only if taken abroad. / Memberships in what but without why. / Honors, but not how they were earned. / Write as if you'd never talked to yourself / and always kept yourself at arm's length. / Pass over in silence your dogs, cats, birds, dusty keepsakes, friends, and dreams. / Price, not worth, / and title, not what's inside. / His shoe size, not where he's off to, / that one you pass off as yourself. / In addition, a photograph with one ear showing. / What matters is its shape, not what it hears. / What is there to hear, anyway? / The clatter of paper shredders.[30]

Even the inauthentic passport of oneself is uncaringly destroyed in the end. The essays collected here also come from the non-academic self, yet the one nurtured at the same springs of the joint muses of poetry and history. It is a different way of seeing the world. A different way of seeing the world that came not only from poetry but also from place. I was impressed by many winter train journeys from East to West Germany, where in snowstorms I was whirled through Breughel-like landscapes of black forests stark against the whiteness. It also came out of many night-time walks through East Berlin, particularly from the Wilhelmstrasse where I worked, up Unter den Linden, onto the Museum Island and over the river Spree and up to the Rosenthalerstrasse, where I lived. In Berlin for the first time, my inner and outer worlds seemed to coincide. In the deep cold of that first winter, something blossomed. I became able to write in a way that fully represented that inner world for the first time. Previously, the inner world had popped willy-nilly into academic writing, like the section on 'inner space' in my essay "New Spaces in Natural History". Insistent concerns in my academic work, like the making and finding of vocation, the entering of the 'life-path,' very much reflecting my own uncertainties as a historian entering a difficult professional world, could be subsumed in this surrender to this place. It was, of course, also a place where atrocious things had happened and had been planned. A place where order and *Gemütlichkeit* had tipped over into terrible opposites. That I should feel so deeply in tune with such a place never ceased to concern me. I still cannot explain it.

In conclusion: It has been a long arc, from Cambridge to Rochester over Australia and Berlin. What has sustained it has been the writing and teaching of history. And, since both are intimate acts, a cogent introspection in divided worlds – a perpetual struggle to break the glass wall that divides us from the past.

Notes

1 Greg Dening, *Islands and Beaches: Discourse on a Silent Land: Marquesas 1774–1880* (Honolulu: University Press of Hawai'i, 1980), 26.
2 Dening, *Islands and Beaches*, 31–32.
3 Marion Milner, *Eternity's Sunrise: A Way of Keeping a Diary* (London: Virago, 1987), 30–31.
4 Dorinda Outram, *Four Fools in the Age of Reason: Laughter, Cruelty and Power in Early Modern Germany* (Charlottesville: University of Virginia Press, 2019).
5 Dorinda Outram, *Panorama of the Enlightenment* (London: Thames and Hudson, 2006).
6 Dorinda Outram, *The Enlightenment* (Cambridge and New York: Cambridge University Press, Fourth edition, 2019).
7 Eliza Lucas (Pinckney) is now celebrated as the pioneer of indigo growing in South Carolina. For quotation, see Hariott Horry Rutledge Ravenel, *Eliza Pinckney* (New York: Scribner's Sons, 1896), 49.
8 David Hume, *A Treatise of Human Nature* (1739) I, iv, 6: "Of personal identity".
9 Enid Welsford, *The Fool: His Social and Literary History* (London: Faber and Faber, 1935).
10 Welsford, *Fool*, 193–194.
11 Wisława Szymborska, *View with a Grain of Sand: Selected Poems* (New York: Harcourt Brace & Company, 1995), 120.
12 Dening, *Islands and Beaches*, 4.
13 Aaron Sachs, "Letters to a Tenured Historian: Imagining History as Creative Non-Fiction – or Maybe Even Poetry", *Rethinking History* 14 (2010), 5–38; 10.
14 Robert Graves, *The Crowning Privilege: A Selection of Lectures and Essays Concerned with Professional Standards in Poetry* (London: Penguin Books, 1959), 119, 128.
15 Ernst Robert Curtius, *European Literature and the Latin Middle Ages* (Princeton, NJ: Princeton University Press, 1990 (1948)), 228–246, 'The Literary History of the Muses'.
16 Helen Waddell, *The Wandering Scholars* (London: Constable & Co., 1927), 160.
17 Curtius, *European Literature*, 597.
18 Waddell, *Scholars*, 180.
19 Waddell, *Scholars*, 162.
20 Waddell, *Scholars*, 43.
21 Waddell, *Scholars*, 146.
22 Waddell, *Scholars*, 11.
23 Helen Waddell, *Peter Abelard: A Novel* (London: Constable & Co., 1933).
24 Monica Blackett, *The Mark of the Maker: A Portrait of Helen Waddell* (London: Constable & Co., 1973), 220–221.

25 Helen Waddell, *The Desert Fathers: Translations from the Latin* (London, Constable & Co., 1987 [1936]).
26 Waddell, *Desert Fathers*, viii.
27 Waddell, *Desert Fathers*, 29, 31.
28 Waddell, *Desert Fathers*, 18, 19–20.
29 Blackett, *Maker,* 83–104.
30 Szymborska, *Grain of Sand,* 155, "Writing a Resumé".

EDUCATION AND POLITICS IN PIEDMONT, 1796–1814

From: *Historical Journal*, 19 (1976), 611–633. © Cambridge University Press. Reprinted with permission.

In 1820 many of the leading figures in the governments of the Italian states were men who had already been prominent before 1796, and had collaborated with the French during the period of the Empire. Vittorio Fossombroni and Neri Corsini in Tuscany[1] and Prospero Balbo in Piedmont[2] are the outstanding examples in the years immediately following the Vienna settlement. The political survival of these men into a Europe dominated by violent reaction against the events of the preceding twenty years poses interesting questions. How had the pre-revolutionary Italian ruling aristocracies reacted to the experience of Napoleonic

---

1 For Fossombroni (1754–1844) and Corsini (1771–1845) see A. Carraresi, 'La politica interna di Vittorio Fossombroni nella Restaurazione', *Archivio Storico Italiano,* cxxix (1971), 267–355.

2 Balbo (1762–1837) was left fatherless in 1765, and adopted by his maternal grandfather, Count Lorenzo Bogino, minister of the Interior. He studied law at the University of Turin, and graduated in 1780. In 1782 he helped to found the *Patria Società Letteraria* and became one of the Decurioni or municipal officials of Turin. In 1788 he became Secretary of the *Accademia delle Scienze*. Between 1796 and 1799 he was Sardinian ambassador in Paris. After Piedmont was invaded by France in 1798, he accompanied the exiled sovereign to Sardinia and Tuscany. He accepted employment under the Austrian regime in Turin in 1800, but resigned his position after a few weeks. In 1802 he returned to Turin, and refused several offers from the French of employment on the Conseil d'Etat in Paris. In 1805, however, he accepted the comparatively modest position of Rector of the University of Turin. In 1816 he was made ambassador to Madrid, and in 1818 placed at the head of the Magistrato della Riforma, or general administration of education in Piedmont. In 1819, he combined this position with that of minister of the Interior, a post which he lost in the aftermath of the revolution of 1821. His critics asserted that these disturbances originated in the liberal tendencies which he had fostered in the University of Turin, and in the dangerous diffusion of knowledge among the lower classes caused by his enthusiasm for the spread of primary education. His last years were spent in scholarly retirement. There is no adequate biography of Balbo, who has been overshadowed by the renown of his second son Cesare. The best modern account is that of F. Sirugo in *Dizionario degli Italiani* (Rome 1963), vol. v. Detailed examination of the events of 1821 is contained in P. Egidi, *I moti studenteschi di Torino nel gennaio 1821* (Turin, 1923). Material relating to his career as Rector of the University is lengthily misinterpreted along nationalist lines by E. Passamonti, 'Prospero Balbo e la rivoluzione del 1821', *Biblioteca di Storia recente,* vol. xii, ed. Rossi and De Magistris (Turin, 1926), 190–347.

DOI: 10.4324/9781003038085-2

government, and how did this experience affect their attitude to the events of the succeeding decade?[3]

The annexed Italian territories were financially vital to the needs of France. Gaudin, Napoleon's Minister of Finance, went so far as to exclaim that the fate of the Empire was settled on the plain of Marengo.[4] Yet precisely because of their importance, the Italian territories also presented France with serious problems, and their leaders could exert great pressure on her. To obtain men and money from Italy through the channel of an efficient bureaucracy, it was essential to reestablish the political harmony shattered by the violent internal struggles which racked the Italian states in the 1790s. Gaudin was careful to continue his appreciation of the importance of Marengo by pointing out that 'la victoire elle-même eut été inutile si l'administration n'avait pas été toute préparée pour en recueillir les fruits'. To capitalize on their victory, the French had to create a civil administration commanding as wide a support as possible; in the event, this was to mean an administration staffed by as many of the Italian nobility as could be induced to serve.

It is well known that the French in Italy quickly abandoned the extremists who had paved the way for their initial victories, and turned to the moderates for support.[5] It was difficult to discipline and exploit the tiny provincial republics which were proclaimed in Piedmont after the first French offensives in 1792. Later, it was just as difficult to control the factional struggles in Turin after the collapse of the monarchy and the full French military occupation of late 1798 to late 1799. Accordingly, the French tended increasingly to rely on the collaboration of relatively docile moderate republicans. However, it is rarely pointed out that, between 1800 and 1802, French policy towards Italian collaborators underwent yet another shift. Serious efforts now had to be made to gain the allegiance of the nobility, the very class which had suffered most in the revolutionary period. The French needed to give their rule an aura of legitimacy and stability which the adherence of moderate republicans, mostly drawn from the middle and professional classes, could never provide. Efficient administration through a stable regime would then become possible.

3 There are surprisingly few studies of the Italians who collaborated with the French regime. This deficiency is especially marked in the case of territories outside the Regno d'Italia. Little advance has in fact been made since the study of Tommaso Corsini, 'Di alcuni cooperatori di Napoleone I', in *Ritratti e studi moderni* (Milan, 1914), 397–459. See also the appeal for more detailed biographical study of these men in J. M. Roberts, *Francesco Melzi d'Eril, an Italian Statesman (1796–1806)* (unpublished Oxford D.Phil. thesis, 1954).

4 Martin-Michel-Charles Gaudin, duc de Gaëte, *Mémoires, Souvenirs, Opinions et Ecrits* (2 vols. Paris, 1826), I, 170–1.

5 P. Gaffarel, *Bonaparte et les Républiques italiennes, 1796–1799* (Paris, 1895), p. 29 and note, quoting Bonaparte's famous letter to the Directory of 28 Dec. 1796, 'il y a en ce moment en Lombardie trois partis: 1. celui qui se laisse conduire par les Français; 2. celui qui voudrait la liberté et montre même son désir avec quelque impatience; 3. le parti ami des Autrichiens et ennemi des Français. Je soutiens et j'encourage le premier, je contiens le second, et je réprime le troisième.'

In this paper I examine the reactions of a group of Piedmontese nobles, gathered round Prospero Balbo, to their involvement with the administration of education under the French. Why did they agree to collaborate in the first place? French promises to hold down the radical groups were probably the most decisive factor. Other possible reasons, such as economic pressure, may be advanced, but cannot be convincingly sustained. It is undoubted that the period between 1796 and 1814 was one of severe economic stress in northern Italy. But the salaries attached to the upper strata of the University administration in Turin, where appointments among the nobility were to be concentrated, were hardly sufficiently high, at a maximum of 6,000 fr. p.a. for the Rectorship, to overcome all scruples of conscience against collaboration with the French.[6]

In the case of Balbo himself, it could also be argued that his diplomatic career had been instrumental in turning his thoughts towards collaboration. After the Austrian victories of 1799, he certainly considered that Piedmont might have a better chance of survival as a buffer-state within the French orbit than as an Austrian protectorate. But the definitive annexation of Piedmont to France in 1800, and her full legal incorporation with the interior in 1802, would render meaningless the considerations of 1799.[7]

The real reasons for the eagerness of sections of the Piedmontese nobility to accept employment under the French are more convincingly to be found in the political conflicts which centred round the control of education in Piedmont between 1800 and 1802. This area of government had assumed a special importance because republican politics in the 1790s had been dominated by university and college teachers, and because universities and schools were widely credited with the diffusion of new ideas emanating from revolutionary France.[8] Under the provisional government of Piedmont by General Jean-Baptiste Jourdan[9] this

---

6 Among the teaching staff of the University, the position was rather different. Salaries, never generous (around 2,000 fr. p.a. for a professor) declined in value as the economic situation worsened. Nevertheless, prospects were no brighter in any other kind of employment, and the dissolution in 1809 of the religious orders, to which many teachers belonged, even further increased their dependence on their university salaries. University administrators and university teachers, proceeding from very different economic bases, were thus faced with very different financial problems.

7 For Balbo's thought on the situation of Piedmont in 1799, see E. Passamonti, 'Una memoriale inedita di Prospero Balbo nel dicembre 1799', *Accademia delle Scienze di Torino, Atti*, XLIX (1913–14), 914–51. J. M. Roberts has argued, in relation to the Austrian subject territories in Italy, that a strong tradition of the acceptance of *de facto* governments made it easier for the Lombard nobility to collaborate with the French. (*Francesco Melzi d'Eril* § 1). However, these considerations obviously do not apply to Piedmont, ruled by a native dynasty.

8 See the social analysis contained in Giorgio Vaccarino, 'L'inchiestà del 1799 sui Giacobini in Piemonte', *Rivista Storica Italiana*, LXXVII (1965), 27–77.

9 Jourdan's regime lasted from 1800 to 1802. There is no adequate account of his career, or of the military republican opposition to Napoleon. For Jourdan's disapproval of the coup of brumaire, see *Correspondance de Napoléon I . . . publiée par ordre de l'Empereur Napoléon III* (32 vols. Paris 1859–), VI, p. 14, no. 4397. Some additional information on military republicanism can be obtained from H. Auréas, *Un Général de Napoléon: Miollis* (Paris, 1961). Miollis was governor-general of the Roman *départements* from 1809 to 1814.

influence remained. His immediate Italian subordinates, making up the Executive Commission of Piedmont, were Carlo Botta,[10] Carlo Bossi[11] and Carlo Giulio. Botta was also a member of the Conseil d'Instruction Publique which took over responsibility for the University of Turin from the Magistrato della Riforma of the old regime. Other members of the Conseil were Sebastiano Giraud and Carlo Brayda. Giraud, like Bossi, Botta and Giulio, was a former member of the staff of the University of Turin.

Under Jourdan's guidance, the Conseil quickly reorganized the University in Turin. The faculties were regrouped and renamed Ecoles Spéciales after the French model. Literary teaching was relegated to the secondary schools, and many professors displaced. Most important of all, the teaching of theology was abolished, a measure which could not fail to arouse great resentment. Meanwhile, Jourdan and his secretary Philippe la Boulinière prepared their own report on the future progress of educational reform.[12] In the fortunes of this plan can be traced the combination of factors which led to the recall of Jourdan, the collapse of the Executive Commission, and the return of conservative groups in Piedmont to power under the French.[13]

At a time when republicanism was becoming increasingly attacked in France, and when the social thinking of the Consulate was becoming increasingly conservative, Jourdan's plan emphasized values of a very different kind. The report acknowledges its debt to the educational scheme for France put forward by Chaptal.[14] This scheme had already been strongly criticized by Napoleon and was eventually to be rejected in favour of the famous report by the future director of education, the chemist Fourcroy. This report established

---

10 The only biography of Botta is still that of Dionisotti (Turin, 1867). Some effort towards a complete listing of his known correspondence was made by Arturo Bersano, Carlo Frati and Carlo Salsotto, in, respectively, *Accademia delle Scienze di Torino, Atti*, xxxvi (1900–1), 969–96; xlvi (1910–11), 12–28; li (1915–16), 717–48.

11 F. Boyer, 'Carlo Bossi et le Piémont', *Rassegna Storica del Risorgimento*, lix (1969), 44–57. There is no biographical study of Carlo Giulio, professor of medicine at the University of Turin, who became Prefect of the Stura in 1804, and died in 1814.

12 Archives Nationales, Paris, F. 17. 1603, Organisation . . ., report of 77 pp. dated 'Premier jour complémentaire, an IX'. All documents cited in this paper are to be found in the Archives Nationales unless otherwise stated. Little information on La Boulinière is available. He seems to have left no printed works. He was dismissed from his post of Secretary of the French administration of Piedmont when Jourdan was recalled in 1802, to reappear in 1804 as professor of geography and history in the University of Turin. I have been unable to discover any information after that date.

13 The importance of issues concerning the University of Turin and its staff in determining the crisis of 1802 has never been fully realised. F. Boyer, 'Les institutions universitaires en Piémont de 1800 à 1802', *Revue d'Histoire moderne et contemporaine*, xvii (1970), 913–17, is an extremely brief summary of the main events.

14 Report quoted, p. 41. Antoine Chaptal (1756–1832) was minister of the Interior from 1800 to 1804. See J. Pigeire, *La vie et l'oeuvre de Chaptal* (Paris, 1932); for his educational thought, R. Tressé, 'J. A. Chaptal et l'enseignement technique de 1800 à 1819', *Revue de l'histoire des sciences*, x (1957), 167–74.

the tightly organised, militaristic, and socially exclusive *lycées* in place of the existing *écoles centrales*.[15]

The Jourdan scheme emphasized a much less restrictive social base for educational planning. To declare that 'tous les citoyens doivent être habilités à l'administration municipale' was to emphasize egalitarian values at a time when it was fast becoming politically disadvantageous to do so. The report is sometimes even couched in language which might be taken as criticism of the contemporary swing to the right in France, and especially of the gradually increasing prominence of the nobility. 'Nous sommes encore loin du pur républicanisme, tant que la considération due aux citoyens illustres sera aveuglement déversée sur leurs fils, leurs parents, leurs amis, tant que les fonctions publiques leur seront indistinctement confiées. Et nous n'avons réellement fait autre chose que (de) changer de dynastie dans la noblesse héréditaire.' Worse followed. To avoid a return to aristocratic society, the report concluded, everyone must be offered 'les moyens de développer les facultés dont la nature lui a fait présent sans égard des distinctions sociales . . . un état républicain peut subsister sans savants, sans érudits, sans métaphysiciens. Il n'existera jamais sans hommes instruits de leurs droits et leurs devoirs, sans citoyens.'[16] Where Fourcroy saw the educational system as a means of reinforcing existing social divisions, Jourdan's view was quite the reverse. A system where the main interest of the state lay in the diffusion of specialised knowledge at *lycée* and university level would have been totally opposed to his philosophy; so also was the restoration of the aristocracy in Piedmont to the position of cultural and governmental authority which they had enjoyed before 1796.

Jourdan's plan was not easily implemented. At the time it was written, it was uncertain whether a single plan for the organization of education would be adopted in all French territories, or whether each region would be allowed to evolve its own. After submitting his plan, Jourdan wrote to the minister of the Interior emphasizing the political benefits to France of a rapid reorganization of education in Piedmont. Without it, the French could not hope to influence the rising generation in their favour. The minister merely replied that he must wait for a decision until the fate of the government's own plan for education in all French territories had been discussed in the *Corps Législatif*. Meanwhile, officials criticized the Jourdan scheme for laying too much emphasis on the provision of primary education, and for attempting to break away from the centralized planning of Paris. The issues at stake were clearly much wider than the future of Piedmontese schools. The degree to which the provinces of the Empire should be integrated with metropolitan France was also under discussion. The centralized

15 L. P. Williams, 'Science, education and the French Revolution', *Isis,* XLIV (1953), 311–30. For Fourcroy, see G. Kersaint, 'Antoine-François de Fourcroy, 1755–1809', *Mémoires du Muséum National d'Histoire Naturelle,* série D, I (1966), 1–296. In his lifetime it was widely believed that he had betrayed his master, Lavoisier, to the guillotine. R. Hahn, 'Fourcroy advocate of Lavoisier?', *Archives Internationales d'Histoire des Sciences,* XII (1959), 285–8.
16 Report quoted, pp. 2, 10, 14.

French state did not spring fully armed from the head of Napoleon; it was the result of several years of bargaining and compromise with both French and Italians interested in the government of the occupied territories.

In the end Jourdan obtained permission to implement his plan only for another year.[17] But twelve months later, discussions in Paris on the foundation of a *lycée* in Turin were already well advanced, and the centralized plan embodied in the Fourcroy report had been approved for use in all territories legally incorporated into metropolitan France.[18] Jourdan began to use Fabian tactics to preserve his scheme. When asked why the *lycée* in Turin was taking so long to open, he emphasized the difficulty of gathering sufficient information to prosecute such an important task; and as long as he remained in Piedmont, information continued to be extraordinarily difficult to gather. Meanwhile, Carlo Botta went to Paris to try to negotiate a favourable decision on the fate of the scheme.[19] But before anything had been decided, Jourdan was recalled to Paris and replaced by a temporary governor-general, Alexis Charbonnière,[20] who controlled Piedmont until the arrival of the new permanent governor, General Menou.

Charbonnière's task was to smooth Menou's passage into Turin. To do this, he had to dispose of the Executive Commission, which had been closely linked to Jourdan and the republicanism shown in his educational plan and his long rearguard action in its defence. After the annexation of Piedmont, the Commission was no longer needed by the French; in fact it became harmful to them, because of the factional disputes aroused in Turin between other political groups jealous of its pre-eminent position. The French could not long afford to tolerate such political instability in an important subject state.[21]

The new envoy launched his attack on the Commission by the dismissal of the Conseil d'Instruction Publique on charges of corruption in the administration of the revenues of the University and of attempting to block the introduction of the French educational system into Piedmont. Charbonnière appointed his own investigating commission to provide evidence to substantiate these charges. The Conseil itself was replaced by a new body[22] composed of Balbo's friend, Count Giovanni

---

17 F. 17. 1603, Organisation . . . fo. 2, 15 vendémiaire an X; fo. 4, 4 frimaire an X; fos. 14–16, n.d.; fo. 26, 9 brumaire an X; fo. 30, n.d.
18 Decree of 21 frimaire, an IX.
19 F. 17. 1603. Organisation . . . fo. 45, 15 vendémiaire an XI.
20 Alexis Charbonnière (1778–1819) began his career as a cavalry officer under the old regime. In 1802 he was appointed secretary-general *per interim* to the French administration in Piedmont, and in 1806 was rewarded for his services by a post in the Imperial guard of honour. He was also the author of several forgotten plays and poems. Contemporary opinion on his political work in Piedmont is contained in Giorgio Vaccarino, 'La classe politica piemontese dopo Marengo, nelle note segrete di Augusto Hus', *Bollettino storicobibliografico subalpino*, LV (1953), 5–74.
21 Report of Charbonnière, F. 1E 78, 13 nivôse, an XI.
22 Reports in F. 17. 1606, Personnel et Affaires diverses. See also Brayda, Botta et Giraud, *Vicissitudes de l'Instruction Publique en Piémont depuis l'an VII jusqu'au mois de ventôse an XI* (Turin, 1803).

Saluzzo, his old teacher Baudisson,[23] and his relative, Count Faletti-Barolo. The investigating commission was headed by Cavalli, the senior magistrate in the Appeal Court in Turin, and a close friend of Baudisson. In this way, a tightly linked group of conservatives dominated the attack on the Conseil and thus precipitated the fall of the Executive Commission, whose members had served on both bodies. The close connection of the members of the Commission and of the Conseil with republican movements within the University in Turin, and their interests in the direction of education after 1800, had determined that Charbonnière's attack would centre round the issue of the control of educational institutions.

The victory of the nobility was fully supported by Menou.[24] Their demands for the preservation of the old educational structure, and for the return of religious teaching and observance to the University were sympathetically received, and led to their willing acceptance of the French system, which was not very different from that operating in mainland Piedmont before 1796. Saluzzo pointed out with relief that 'les lycées sont donc assez conformes à nos collèges d'instruction provinciale, et il n'y a d'autre différence qu'en ce qui se rapporte aux enseignements de la langue française'. His sentiments were echoed in Paris, where it was emphasized that 'en donnant à Turin un Lycée et quelques écoles spéciales, aux villes inférieures des écoles secondaires, et aux communes des campagnes des écoles primaires, les choses se trouvent à cet égard dans le Piémont à peu près sur le pied où ils étoient sous l'ancien gouvernement'.

Preparations for the opening of the *lycée* suddenly began to go ahead with considerably greater speed than they had under Jourdan. Inspectors sent out from Paris praised the co-operation and efficiency of the new governor-general, just as they praised the political means which he used to obtain this administrative efficiency. They well understood the importance of that 'calme que l'Administrateur-Général a contribué à établir par tous les moyens de rapprochement qui sont dans sa main'.

It was thus on a basis of cooperation between the French and the Piedmontese nobility that the introduction of the French system of education was effected.[25]

23 See I. M. Baudisson, *Orationes pro Comite Prospero Balbo* . . . (Turin, 1780). Baudisson, a professor of canon law, had lost his chair in the reorganization of the University by the Conseil d'Instruction publique.

24 F. 17. 1603, Correspondance . . . an XI. No study exists of Menou's Italian career. Of aristocratic origins, he was a loyal supporter of Napoleon, and accompanied him on the Egyptian campaign. After the assassination of Kléber, and the departure of Napoleon, he became commander-in-chief of the army of the Orient. In 1808 he was made governor of the newly annexed Tuscan departments, and in 1810 governor of Venetia. He died in 1812. Paul Marmottan's *Le Général Menou en Toscane* (Paris, 1904) briefly sketches a few of the more lurid episodes of his career. His extensive correspondence with Eugène Beauharnais, Governor of the Regno d'Italia, is preserved in Princeton University Library.

25 F. 17. 1603, Correspondance . . . an XI, 'Délibération du Conseil-Général du Département du Pô, 6 floréal, an XIII . . . sur la proposition du C. Baudisson, de demander . . . la conservation de l'Université de Turin'. For Saluzzo's comments, see F. 17. 1607, Personnel; for the comments of the Inspectors Villars and Lefèvre-Ginau, see F. 17. 1611, Lycée, report of 25 frimaire an XII.

But the nobility were to discover that collaboration with the French, however necessary, was neither pleasant nor easy. Their political accommodation with the occupying power could only arouse intense resentment among the moderate republicans now relegated, after a brief taste of power, to the class-room and lecture-hall. Further, the ordinary administration of institutions in the annexed territories was full of painful surprises, especially in relation to financial questions.

The mechanisms are now well understood by which the French liquidated the state debts of the occupied territories, invested the funds so gained for their own profit, and at the same time funded the remaining institutions in these territories from such investments.[26] However, the internal functioning of institutions funded by these methods has received little attention. In the case of the universities in Italy, Napoleon's own statements of policy clearly stated that the universities would as far as possible finance themselves from the interest on their endowments, which the French had invested on their behalf in state loans. Direct subsidies to the universities from government departments were frowned upon.[27] This scheme could only function efficiently given an adequate initial endowment, and a measure of promptitude and honesty on the part of the agencies charged with the payment to the University of its interest. In the case of Turin, these provisos were not met. The university, enormously wealthy before 1796, now suffered a crippling financial burden,[28] and the conduct of the French financial agencies strained the loyalty of Balbo and his colleagues almost to breaking point.

Between 1799 and 1805, the University had administered its own endowments, which provided it with an annual income of about 500,000 fr. Such revenues and such independence could not last long. Very early on, Saluzzo expressed his fears that the university would be placed at the mercy of 'la cupidité des financiers'.[29] His forebodings were justified. On Napoleon's initiative, the minister of the Interior had already decided the university's fate.[30] The Domaine sold the university's property and invested the proceeds in 5% consolidated shares. From this, it was to pay the University annually the resulting 300,000 fr. The Domaine's profit came from delaying payment of the interest and reinvesting it during the delay.

---

26 Marcel Marion, *Histoire Financière de la France depuis 1715* (6 vols. Paris, 1927), vol. IV, ch. 6, 8–9.

27 *Correspondance* . . . XVIII, 89, no. 14503; F. 17. 1602, fo. 233, doss. 4.

28 Financial difficulties forced Balbo to suspend the opening of the departments of art, pharmacy, and veterinary medicine. No salaries could be paid to the professors of music, the keeper of the botanical gardens, or to the staff of the Museum of Natural History. It was impossible to open a new anatomy theatre, establish an extra chair of medicine, or a new student hostel, all of which improvements had been envisaged by the 'reorganisation' of 1802. Salaries of teaching staff were paid at least three, and often nine, months in arrears. F. 17. 1605, letter of 24 Apr. 1807; F. 17. 1603, report of 28 Apr. 1811.

29 F. 17. 1607, Personnel et notes diverses, Saluzzo to Fourcroy, 12 messidor an XI.

30 *Correspondance* . . . VIII, 278, no. 6593; IX, 651, no. 8008. F. 17. 1607, ibid. minister of the Interior to minister of Finance, 13 prairial XIII; F. 17. 1603, fo. 36, 18 germinal XI.

Furthermore, of the 300,000 fr., one-tenth was to be retained by the University as an endowment fund, so that its real annual revenue was 270,000 fr., little more than half the amount obtained before 1805.

Difficulties over payment through the Domaine began almost immediately. Delays of up to ten months were common, and left the university with an initial deficit of 157,607 fr.[31] When payment was eventually obtained, it was in the form of drafts on the Receveur des Domaines in Turin, and on the Receveur Général for the *départment*.[32] This meant that the university would again be faced with delays in payment, since the Receveurs also depended for their profit on prolonging the period between the receipt and payment of funds so that they could speculate with the money they held.

Other drafts, on the Caisse d'Amortissement, were also given to the university. This was just as likely to delay payment, as the Caisse was regularly used by Napoleon to maintain the levels of purchase of state loans at times when confidence in the regime was low. When levels of purchase were high, interest payments tended to be reduced, and no exception was made for the University.

In 1811, the atmosphere became increasingly heated, and Balbo and the governing body of the University accused the Domaine of swindling them by misrepresenting the amount of revenue obtainable from the University's endowment. In August, part of the missing sum was ordered to be paid, but no cash ever materialized, and unpaid arrears continued to cripple the University's finely-balanced finances. By 1813, functionaries of the central educational administration in Paris regarded their defeat at the hands of the financial agencies as inevitable. Attached to one of Balbo's letters of complaint is a despairing comment:

Voici une note de M. de Balbe sur la manière dont l'Académie de Turin est payée des rentes qui lui appartiennent. Comme l'expérience m'a prouvée que toute réclamation relative aux finances est inutile et de nul effet, je me borne à vous transmettre cette note, sans y joindre aucune observation, et seulement pour l'acquit de ma conscience.[33]

Financial transactions such as these, unpleasant, frustrating, and with disturbing implications for long-range planning, made up the greater part of the day-to-day administration of the University. This renders even more pressing the need to explain what made continued collaboration worthwhile for the Italians.

The origin of their reaction must be sought at a period far earlier than that of the annexation of 1802. Nearly half a century before, in 1759, the young Count

---

31 F. 17. 1609, Ecole de Médicine, procès-verbal of 29 brumaire XIV; F. 17. 1607, letter of 12 May 1807.

32 F. 17. 1613, Domaines, letter of 1 June 1813.

33 Ibid. note of Coiffier to Cuvier, 30 Aug. 1813. See also F. 17. 1605, procès-verbal of 11 July 1811, para 51. 'Le Directeur du Domaine du Pô à force de subterfuges et de prétextes évasifs a réussi jusqu'à présent à contraver et rendre inutile toute espèce de démarche'.

Giovanni Saluzzo, Count Giovanni Cigna[34] and Louis Lagrange, the famous mathematician, had together founded the *Accademia delle Scienze* in Turin. They soon attracted royal patronage and began to publish the scientific work of many famous scholars, especially those of French nationality. In 1788 Balbo, who was also involved with many other newly-founded learned societies, became secretary to the *Accademia*. In this way, he and his friends became part of the contemporary European revival of interest in learned societies.[35] They also placed themselves in contact with the leading figures of the French scientific world, many of whom were to be metamorphosed by the Revolution into propounders and administrators of the new state educational organisation. Condorcet contributed papers to the proceedings of the *Accademia*; Lavoisier corresponded with it. Saluzzo in later years referred proudly to his 'assez constante correspondance' with both men.

Balbo's involvement with the revival of the academies would have allowed him to view with sympathy the ideas put forward by Condorcet, during the revolutionary period, on the role of the state in education. Chief among them was the direction of instruction and research at all levels by a master-academy,[36] not very different either from the Imperial University established by Napoleon between 1806 and 1808, or from the University of Turin as it was in Balbo's youth. Reforms undertaken in Piedmont in 1772 had given the University, governed by the *Magistrato della Riforma*, the surveillance of all stages of education in mainland Piedmont. It was often claimed later, with some plausibility, that the University of Turin had inspired the pattern of the Imperial University. It is certainly true that in Piedmont the manipulation of educational institutions to produce docile citizens was received practice long before 1796.[37] Balbo and his group were thus well prepared by their own experience, as well as by their French contacts, for the form that the organization of education was to take under Napoleon.

Actual attendance at the University of Turin also created close personal links between men who were to follow very different paths in the 1790s. Michele

---

34 For Cigna's contribution to the theory of combustion, see R. Fric (ed.) *Oeuvres de Lavoisier . . . Correspondance* (3 vols. Paris, 1953), II, 432–3, 461.

35 G. Torcellan, 'La Società Agraria di Torino', *Rivista Storica Italiana*, LXXVI (1964), 530–52; E. W. Cochrane, *Tradition and Enlightenment in the Tuscan Academies, 1690–1800* (Rome, 1961); Roger Hahn, *The Anatomy of a Scientific Institution: the Paris Academy of Sciences, 1666–1803* (Berkeley, California, 1971).

36 K. M. Baker, 'Les débuts de Condorcet au Secrétariat de l'Académie Royale des Sciences, 1773–1776', *Revue d'Histoire des Sciences*, XX (1967), 229–80. For the continuing acceptance of this idea in Italy, see *Magasin Encyclopédique*, XL (1801), 96, 'Le C. Cagnoli, Président de la Société Italienne des Sciences au C. Delambre, Secrétaire de l'Institut National à Paris, Lyon, 3 pluviôse an X'.

37 E. Rendu, *A. Rendu et l'Université de France* (Paris, 1861), p. 42; M. Viara, 'Gli ordinamenti della Università di Torino nel secolo XVIII', *Bollettino Storico-Bibliographico Subalpino*, XLV (1942), 42–54. For Bogino's policy towards the University of Cagliari, see F. Venturi, 'Il conte Bogino, il Dottor [sic] Cossu, e i Monti Frumentari', *Rivista Storica Italiana*, LXXVI (1964), 470–506. For earlier policy towards educational reform in Piedmont, see G. Quazza, *Le riforme in Piemonte nella prima metà del settecento* (2 vols. Modena, 1957) II, ch. X.

Brugnone, professor of medicine, and member of the government of 1798; Carlo Giulio, professor of medicine and member of the Executive Commission of 1800; and Benedetto Bonvicino, professor of medicine and president of the Municipality of Turin in 1801, all frequented Balbo's group, and were members of the *Accademia*.[38] The diversity of the political views adopted by those who had been so intellectually united in the 1770s and 1780s should perhaps lead to some modification of the now fashionable interpretation of the revolutionary period in terms which equate radicalism with an interest in the new physical and social sciences emerging at the end of the century.[39]

Under the Empire, the staff of the University of Turin in fact represented the whole spectrum of political outlook. Some had supported the union of an independent Piedmont with the Cisalpine Republic, like the Boyer brothers, professors of hydraulics and mechanics. There were also those who, at first sympathetic to the French cause, had not dared brave prison or exile in 1799, and had recanted their opinions. Filippo Regis, professor of law, though president of the *Società Patriotica* in Turin, yet published a long declaration of monarchist principles after the Austro-Russian victory, thereby escaping the penalties imposed on extreme republicans such as the medical professors Balbis and Buniva.[40] Others collaborated with the French, but wisely chose to do so after political confusions had been resolved by the purge of the Executive Commission in 1802. Antonio Franchi, another professor of law, for example, had acted as secretary to the Cavalli commission which provided material for the attack on Botta, Brayda and Giraud. After their fall, his career went from strength to strength. In 1804 he was a member of the electoral college for Turin, and a candidate for the *Corps Législatif* in Paris.[41] Another professor of law, Gaspare de' Gregori, resisted pressure from the extreme radical Ranza to join the annexationist party. This piece of political wisdom earned him a sub-prefecture at Lanzo in 1801. In 1809, he was deputy in the *Corps Législatif* for the Sésia, and in 1811, together with Cavalli, was given high legal office in Rome.[42]

It would seem that among the teaching staff of the university the French had been successful in creating an amalgam of all political parties, prepared to work together under the French. But a closer examination shows that this was far from the case. Real extremists had been completely excluded. Of the rest, it was rare for any but the willing collaborators, with records clean of political involvements in

38 Most of these details are to be found in Prospero Balbo, *Lezione accademiche . . . intorno alla storia della Università di Torino* (6 vols. Turin, 1825) II, 194ff. Contemporary scientific work in the University is described in A. Pace, *Benjamin Franklin and Italy* (Philadelphia, 1958), ch. 'Erupuit Caelo Fulmen'.
39 E.g. R. Darnton, *Mesmerism and the End of the Enlightenment in France* (Cambridge, Mass., 1968).
40 N. Bianchi, *Storia della Monarchia Piemontese dal 1779 sino al 1861* (4 vols. Rome–Florence–Turin, 1877), III, 339; F. 17. 1607, Faculté des Sciences, Etats de service.
41 F. 17. 1607, Ecole de Droit, Balbo to Champagny, 2 Apr. 1807.
42 A. Bersano, 'Un conformista: Gaspare Antonio de'Gregori,' *Bollettino Storico-Bibliografico Subalpino,* LXVI (1968), 523–40.

the last decade, to be accorded honours by the French. Careful police supervision of the University was in any case always on hand to repress movements of discontent.[43] No real reconciliation had taken place. The remaining republicans had been left powerless, but discontented at the favour shown to the nobility by Napoleon.

If the policy of *amalgame* so often enunciated by the emperor[44] had really been intended to be put into operation, Balbo's career in the period before 1806 should have disqualified him from holding office in an administration so filled with disgruntled republicans. Few could have shown themselves more faithful to the Sardinian monarchy. As Piedmontese ambassador in Paris, he tried unsuccessfully between 1796 and 1799 to avert the advance of the French. When the monarchy collapsed, he followed the king into exile in Sardinia and Tuscany, only returning to Turin when in 1802 all émigrés were threatened with the confiscation of their property. He remained completely apart from public life until his appointment as rector of the University in 1805, refusing several offers of employment in the Conseil d'Etat in Paris. In this interval of private life, he and a group of friends decided to attend to the education of their children, rather than expose them to the baneful influence of the French institutions. Saverio Provana and Filippo Grimaldi taught the children mathematics and physics, and Balbo himself took charge of their literary studies. Maintenance of the traditions of monarchist Piedmont was also actively taken up by the children themselves. In 1804 a group which included Balbo's second son, Cesare, and his friend, the young Carlo Vidua, founded the *Accademia de' Concordi* and admitted Balbo and his circle as senior members. Its demand for the increasing study of the Italian language as a gesture of protest against French domination cannot have been displeasing to their parents.[45]

If any real policy of *amalgame* had been held by the French such a pronounced royalist as Balbo could never have been appointed as rector of a university, many of whose staff held strongly opposed political views. Indeed, Balbo was even chosen *in preference* to Giovanni Saluzzo, who was far less associated with adherence

43 E. d'Hauterive, *La police secrète du premier Empire: Bulletins quotidiens addressés par Fouché à l'Empereur: d'après les documents originaux inédits déposés aux Archives Nationales* (4 vols. Paris, 1922).

44 For statements of the policy of representing all shades of political opinion in the filling of positions, see *Correspondance* . . . VI, p. 152, no. 5528; VIII, p. 389, no. 6735; X, p. 451, 8663.

45 E. Passerin d'Entrèves, *La giovinezza di Cesare Balbo* (Rome, 1940), p. 9; G. Gentile, 'L'eredità di Vittorio Alfieri', *Opere Complete* (Florence, 1963), vol. XVII; V. Cian, *Gli alfieriani-foscoliani piemontesi ed il romanticismo lombardo-piemontese del primo risorgimento* (Rome, 1934). After 1802 the official language of Piedmont was French. This made the political significance of the language question much greater in the former kingdom than it was in areas such as Tuscany, where Italian was kept as the language of official transactions. It seems likely that Balbo would have used both languages in the domestic circle, especially as his second wife, Madeleine des Isnards, widow of the Comte de Séguin, was French.

to the fallen monarchy. Only a shift in French policy towards a definite favouring of the nobility can explain such an appointment.[46]

The detailed account of Balbo's reception of his nomination to the post must be reconstructed almost entirely from his own account of the matter. Neither Lecestre nor the authorised version of Napoleon's correspondence contains much material on the Italian journey of 1804–5, and the recent monumental edition of Cambacérès' correspondence with the Emperor is similarly deficient.[47] We know that Balbo did meet Napoleon in 1805.[48] The rest of the story is contained in a letter which he wrote to Carlo Emanuele IV, after the Restoration, in order to excuse his collaboration with the invader.

> In the end, all I hoped for was to be forgotten; but in the autumn of 1805 I saw in the newspapers that I had been nominated Rector of the University. This position could have pleased me, because it was neither lucrative nor prominent, because it suited my interests, and above all, because it would save me from other harassment. Nonetheless I remained doubtful (whether I should accept it) and I asked advice from persons loyal to your Majesty and lovers of their country. They all urged me to accept, pointing out that though I might not be able to do all the good I wished to do, I could at least prevent much harm; and that in any case, that I should not leave to some Frenchman chosen by intrigue in Paris and a stranger to our customs, or to some Piedmontese upstart from the revolutionary years, a position so important for the education of our rising generation. These considerations convinced me, and I have nothing to repent of, for my success outstripped my hopes. I laboured long and hard, and suffered bitter conflicts, terrible persecution, and unceasing worries. However, in the end I succeeded in controlling my enemies . . . the old institutions were preserved and, where possible, restored; the old traditions, most glorious for the royal family, were recalled on every occasion, and commemorated, not without daring.[49]

Possible modifications may be suggested in this account. Balbo had frequently been solicited by the French authorities in Turin to accept employment,

---

46 Significantly, Botta ascribed his failure to gain the Rectorship himself to the fears of the French 'de déplaire à la noblesse piémontaise' A. Bersano, 'Il fondo Rigoletti dell'epistolario Botta', Bollettino storico-bibliografico subalpino, LVI (1958), 351–79; no. 221. See also A. Neri, 'Una lettera apologetica di Carlo Botta', Archivio Storico Italiano, 5 ser., IX (1892), 76–87.

47 J. Tulard, (ed.) Cambacérès: Lettres inédites à Napoléon (2 vols. Paris, 1974).

48 Bianchi, Storia . . . IV, 369.

49 Quoted in E. Passamonti, 'Prospero Balbo e la Rivoluzione . . .; pp. 206–9. Another version of this letter, in French, is printed in M. degli Alberti, Lettere inedite di Carlo Emanuele IV . . . ed altre, 1814–1824 (Turin, 1900), pp. 30–8.

and the renewed request of 1805 can hardly have come as a great surprise. His years in Paris would also have brought him into contact with many Frenchmen now influential in Italian affairs, including Napoleon himself. While in Paris as ambassador, he had also been called upon to participate in the deliberations of the Commission meeting to determine the bases of the metric system.[50] Here he would have met men such as Louis Lefèvre-Ginau[51] who were later to be instrumental in introducing the French educational system into Piedmont, and who would have served in the intervening years to keep Balbo's name before the French authorities.

However, examination of Balbo's actions as rector does largely bear out his emphasis to Carlo Emanuele on the conservative and defensive aspects of his tenure in office. In the university, Balbo was surrounded by the same group of nobles who had returned to power after 1802, and who had domi-nated intellectual life in Turin in the period before the Revolution. Giovanni Saluzzo, who had also been considered by the French for the post of rec-tor, became Balbo's immediate subordinate, as inspector of the Academy of Turin. Cesare Saluzzo, his son, became a second inspector of the academy by a decree of 27 April 1811. His brother Alessandro became head of the *lycée* in Turin in January 1814. Balbo's uncle, the Abate Incisa-Beccaria, formerly a member of the *Magistrato della Riforma*, returned as head of the Pensionat Académique, or student accommodation; both he and his predecessor in the post, Carlo Adami, had held corresponding positions under the monarchy. In March 1811, Incisa-Beccaria was made a member of the governing body of the University, and was decorated with the title of *Officier* of the Imperial University.

A similar tendency in the making of appointments may be discerned further down the educational ladder, in the Piedmontese *Lycées*. At Casale, Iacintho Carena, another elderly man with a long history of involvement with the Mon-archy, was made *Proviseur*. A friend of Giovanni Saluzzo, he had been for many years secretary to the king's personal office. The financial administrator at Casale was Joseph Pieroleri, a client of Balbo ever since he had been his secretary in the embassy to Paris. Largely through Balbo's influence, Joseph's brother Fran-cesco became secretary to the University in Turin. In running the administration of education in Piedmont, the French showed not only their intention of favour-ing the nobility and their clients; they also revealed their dependence on the

---

50 For the Commission des Poids et Mesures, see M. P. Crosland, *Science in France in the Revolu-tionary Era, described by T. Bugge* (Cambridge, Mass., 1969). Bugge was the Danish delegate to the Commission.

51 Louis Lefèvre-Ginau (1751–1829), mathematician and engineer, member of the Institut in 1795, Inspector-General of Public Instruction in 1802. In 1807 he became a member of the *Corps Légis-latif.*

administrative expertise of the old regime governments to carry on the business of their over-extended Empire.[52]

In the period before the definitive establishment of the Imperial University in 1808, the main development in the life of the university in Turin was the resolution in the favour of the conservatives of the struggle over religious observance. The way in which the university had been reorganized by the Executive Commission had made this issue one of great symbolic importance. A victory for religion was a victory against the radicals. When the university church was reopened in 1807, Balbo commented, 'Tout annonce que . . . les exercices de religion qui auront lieu tous les jours de fête seront très suivis par les élèves, malgré l'opposition de quelques personnes qui heureusement ont perdu beaucoup de leur influence.' In 1810, the teaching of theology was at last restored to the university. Balbo viewed this as a return to the proper traditions of the university distorted in the radical period. He commented to the governing body, 'Rappelée dans notre sein, la faculté de théologie y est accueillie avec d'autant plus de plaisir qu'elle a formé, il y a plus de quatre siècles, le premier noyau de l'Université de Turin.' Shortly afterwards, Balbo went even further in his efforts to preserve university traditions, by opening a library containing 'Une collection des ouvrages qui peuvent servir à l'histoire de l'Université de Turin et de ses anciens professeurs'.[53]

At the end of the period between 1806 and 1808, legal changes took place which should completely have altered the status of the University of Turin. In 1806, the first organizing decrees of the Imperial University were promulgated. This state organization was given complete control over all levels of education, apart from some restrictions on its control of seminary teaching. The territory of the Empire was divided into Academies usually corresponding to the jurisdictional area of the Appeal Court for the same area. The university of that area was, under the control of the central administration in Paris, given the surveillance of all levels of education within the area of the academy. Thus Balbo as rector was responsible for the administration not only of the university itself, but also for primary and secondary education within almost the whole area of the old Monarchy. At the same time he was expected to enforce a complex system of regulations dealing with the new French university examination system, and a whole series of financial

---

52 Similar remarks could be made about the composition of the administration of metropolitan France. Both Gaudin and Lebrun, the financial expert, Third Consul and *Architrésorier* of the Empire, for example, had already achieved distinction in the government of the old regime before 1792. Exploration of this neglected topic would tell us much about the continuities between Napoleonic and royal France. For the rest of the information in the preceding paragraphs, see F. 17. 1607, Personnel, fo. 98, 8 Jan. 1814, para, 1; F. 17. 1611, Lycée, report of 9 Apr. 1811; F. 17. 1604, Personnel, fo. 84; F. 17. 1605, 18 July 1811; F. 17. 1611, Lycée, letter of 13 Feb. 1813.

53 For the theology faculty, see E. Passamonti, *Prospero Balbo e la rivoluzione . . .*, p. 205; F. 17. 1605, Balbo to Champagny, 22 Apr. 1807. For the library, see F. 17. 1613, budget for 1809. This library probably provided much material for the *Lezioni Accademiche*.

transactions between Paris and Turin arising from this system. Apart from this, the new functions of the university were little different from those it performed under the old *Magistrato della Riforma*. After the financial changes of 1804–5, however, the university was left with little more than half of the revenues even of the revolutionary period with which to carry out these demanding functions. It was hardly surprising that Balbo, even apart from his hostility to the French regime as such, should have tried to delay the implementation of the decrees of 1806. He argued that the University regulations of 1806 and those of 1802, formulated by Saluzzo, contained nothing contrary to each other, and that the older regulations should therefore remain in force. For two years he argued the case with Paris, and took care that nothing changed in Turin. In 1808, however, the final organizing decrees of the Imperial University were published. Fourcroy resigned as director of public instruction to be replaced by Louis de Fontanes as grand-master of the Imperial University. The new and more urgent threat brought out an even greater mastery of the tactics of delay on Balbo's part. Those professors in the university who hastened to comply with the new regulations were fobbed off so effectively that one of the professors of medicine wrote a letter of complaint to Paris. Balbo waited six months to react to the charges. When at last he did answer the queries from Paris, he admitted the delays politely, but hardly accounted for them satisfactorily.

> Je supplie votre Excellence de vouloir bien excuser les délais qui ont eu lieu jusqu'au présent, et je puis l'assurer que la presque totalité des agents de l'instruction publique actuellement en place dans l'arrondissement de l'Académie de Turin désire de se mettre parfaitement en règle pour pouvoir [faire] parti de l'Université Impériale.

Strategic delay, however, continued to be used to good effect, and it was only in 1812 that the governing body of the University at last got round to discussing the implementation of many of the basic provisions of the French system. Until 1808, the astonishing weakness of direction from Paris helped Balbo to maintain this polite disregard of French attempts to remodel the university. He made full use of his privilege of corresponding directly with the minister of the Interior, to put his grievances before the highest authority in the most favourable light. At the same time this manoeuvre kept Fourcroy so short of information on the state of affairs in Turin that effective control from Paris was impossible. He admitted, in reply to Balbo's letter, 'Je n'ai pu me former aucun avis sur les objets énoncés parceque je ne connais pas le montant des rétributions ni même l'organisation de l'Université.' Against this background of ineffectual complaint, the dark warnings issued to Fourcroy that Balbo's administration was 'tendant évidemment à ramener l'ancien régime de l'Université dans toute son étendue . . . et à rétablir une instruction publique absolument piémontaise', were unlikely to produce much immediate reaction.

Balbo also continued to fight to retain control of the making of appointments within the University. Only after a protracted struggle did he concede that the right to ratify appointments lay with the grand-master in Paris. Even after this, the making of appointments was in practice still controlled firmly by Balbo and his friends. Many of his clients and relatives had already been found employment in the educational administration. Over the years, Turin eroded the initiative of Paris still further. The manoeuvres leading up to Alessandro Saluzzo's appointment to the *lycée* in Turin in 1814 form a case in point. Balbo described them in a letter to Fontanes.

> Monsieur Hannibal de Saluces, un des écuyers de sa Majesté l'Empereur, a eu l'honneur de parler à votre Excellence d'un projet d'après lequel M. Alexandre son frère, succéderait à M. Adami dans la place de Proviseur du Lycée de Turin ... il jouit dans ce pays de toute la considération qui est due à son nom, à sa mérite, à ses services militaires, et aux connoissances qu'il possède, et qui sont pour ainsi dire héréditaires dans cette branche de son illustre famille.[54]

Balbo's claims that he had as far as possible maintained the traditions, possessions and institutions of the University are thus largely borne out by his actions. The French paid a high price for the politically stabilising effects of reinstating the conservative nobility. Balbo's first loyalty was always to his friends and to the institution he directed, rather than to the French. While the French could weaken the University financially, they could not impose many structural changes without arousing a highly effective opposition.

Further, the central administration of education established in 1808 possessed a highly distinctive character. Handpicked by Napoleon to symbolize the new stable, conservative state he had created in France, the mentality of the officials in Paris played into the hands of the collaborators in Turin. The establishment of the Imperial University completed the movement which the signing of the Concordat and the publication of the *Génie du Christianisme* had begun in 1802. Its functionaries tended to be men who had made their way in the world by supporting Napoleon's campaign against *idéologie*, and in favour of the return of conservative values to public life and of state imperatives to public education. This was the logical consequence of the idea that the untrammelled diffusion of knowledge had caused the horrors and catastrophes of the Revolution. At the head of the university was placed Louis de Fontanes, a man who supported moderate royalism in 1793 and had been exiled in England for his pains, together with his friend Chateaubriand. On his return he collaborated with Lucien Bonaparte on the famous pamphlet, *Parallèle entre César, Cromwell et Bonaparte*, published

---

54 For Fourcroy, and the warnings of the Inspector-General Sédillez, see F. 17. 1703, letters of 21 June 1809, 2 Nov. 1809; For Saluzzo, see F. 17. 1606, letter of 31 Jan. 1812.

in November 1800, but almost immediately suppressed as showing too clearly the First Consul's ambition for supreme power. In 1802 he launched a series of attacks in the press on the liberal opposition, and on Mme de Staël in particular. The campaign culminated in a laudatory review of the *Génie* in the *Mercure* of 15 April 1802. Such was the review's success as a justification of the regime's progression towards conservatism, that Napoleon ordered it to be reprinted in the official *Moniteur* three days later, to coincide with the celebrations which marked the reconciliation of Church and State.

It seemed only fitting that those who through the press had guided opinion towards acceptance of Napoleon's undivided power before 1808, should do so through the medium of educational institutions after that date. Most of Fontanes' assistants shared his background and views. The brothers Philibert and François Guéneau de Mussy, for example, had both been expelled from the Ecole Normale in 1793 for refusing to take an oath of hatred towards monarchies. Under Napoleon, François returned to the Ecole Normale as its Director, and Philibert became an inspector-general of Education. Antoine Arnault, one of the most scathing critics of the Jourdan plan, was equally close to Fontanes; in his autobiography, he stresses the similarity of their views on education, and of their admiration for the Emperor. Previously, Napoleon had used Arnault, a successful dramatist, to strengthen his links with the literary world. Another inspector-general, Jean Joubert, was a friend of Chateaubriand, and Fontanes' collaborator in the article on Mme. de Staël.[55]

After 1808 Fontanes' political views went even further than Napoleon had intended them to do. Support of the Concordat was one thing; support of ultramontanism in the state educational system was another. In 1810 the Emperor's mounting distrust of Fontanes' religious policy came to a head in the famous police enquiry into the conduct of the Imperial University in that year.[56]

Thus the administration of education which appeared in Piedmont after 1808 was moulded by men holding opinions tending to be even more conservative than those of the Emperor himself. These men worked with the conservative groups restored by the French. One of the most important political functions of the Imperial University in the annexed territories was to preserve and stabilize this body of conservative support. Its task was made easier by its substantial agreement on

55 Such examples could be extended. See A. Aulard, *Napoléon I et le Monopole Universitaire* (Paris, 1911); P. Jeannin, 'Une lettre d'Augustin Périer, sur la suppression del'Ecole Normale', *Revue d'histoire moderne et contemporaine*, xv (1968), 466–70; A. Arnault, *Souvenirs d'un Sexagénaire* (4 vols. Paris, 1835), iii, 292; P. de Reynal, *Les correspondants de Joubert* (Paris, 1883), p. 77; A. Wilson, *Fontanes, essai biographique et littéraire, 1757–1821* (Paris, 1928).

56 Ch. Schmidt, *La réforme de l'Université Impériale en 1811* (Paris, 1905). For Fontanes' hostility to Fouché, see L. Madelin, *Fouché, 1759–1820* (2 vols. Paris, 1901). Fouché had been one of the terrorists sent out in 1793 to quell the rebellion in Lyons, Fontanes' native city.

social policy with the attitudes of the collaborators: this made a system workable which otherwise would not have withstood the intense resentment generated by the actions of the financial bodies.[57]

But this very degree of political collusion meant that the French retained only a limited freedom of manoeuvre. Balbo spoke in his letter to Carlo Emanuele of the 'aspri combattimenti e terribile persecuzione' which his functions as rector had compelled him to suffer. Undoubtedly, he was referring at least partly to his prolonged struggle with the medical faculty within the University of Turin. Balbo's delays in introducing the French system of examination were the ostensible causes of the dispute. It is also possible, however, to view it as a movement of political opposition against the conservative monarchist Balbo, by men who had been deeply involved in radical politics before 1802.

The dean of the faculty, Benedetto Bonvicino, had an impressive record of support for the early revolutionary regimes in Piedmont, as

> Commissaire du Gouvernement pour installer les autorités civiles, politiques et militaires dans le département de la Stura, et Président permanent de la Municipalité de Turin en 1799, membre de la Consulte en 1800, de la Commission Municipale . . . en 1801, Président du Collège électoral de l'arrondissement de Coni, et membre du Corps Législatif en 1801.

The background of his supporter Balbis, professor of botany within the medical faculty, showed an even more extreme involvement. In 1798 he had become a member of the provisional government set up in Piedmont by General Joubert. As a friend of Carlo Botta, he was sent out in 1799 to the province of Saluzzo to persuade the municipalities to vote for union with France. During the brief period of Austro-Russian rule, he was denounced as a French supporter and imprisoned. Such was the strength of his reputation as a political agitator that Balbo decided to transfer his chair to the Faculty of sciences, remarking, 'Il ne faut perdre aucune occasion d'affaiblir dans la faculté de médecine l'esprit de faction qui l'agite, malheureux reste des troubles politiques.'[58] Another of Bonvicino's supporters

---

57 The French also made misguided attempts to fix the loyalties of the Italian nobility by a form of educational conscription. The children of noble families were drafted willy-nilly into military schools in the interior of France, or made to accept subordinate positions on the Conseil d'Etat in Paris, as a preparation for office in the administration of the Empire. Naturally, this policy produced precisely the contrary to the desired result in the families concerned. Balbo himself suffered badly. His eldest son, Ferdinand, was forcibly drafted into the army after a spell at St Cyr, and killed on the retreat from Moscow. His second son, Cesare, was pushed into a subordinate position in the French administration in Rome. For the revulsion with which his duties inspired him after the fall of the Papacy, see his *Sommario della Storia d'Italia . . . con autobiografia dell'autore* (Lausanne, 1846).

58 F. 17. 1606, 31 Jan. 1812. Bianchi, *Storia . . .* III, 99, 331; F. 17. 1607, Personnel et Affaires, Etat de Services de J-B Balbis; ibid. letter of 6 Aug. 1809.

was Michele Buniva, famous as the introducer into Piedmont of vaccination against smallpox. In 1799 he led a demonstration mounted by the university in favour of the union of Piedmont with France, and was forced to leave Piedmont after the Austro-Russian invasion. He went to Paris to continue his studies, and came into contact with Antoine Fourcroy. In 1800 he returned to Piedmont and recommenced his political career, which ran closely parallel to that of Bonvicino. As 'membre et secrétaire du comité général du département du Pô . . . Président permanent de la Commission Municipale de Turin . . . Adjoint au Maire de Turin . . . Membre du Collège électoral du département du Pô', he too occupied a prominent position in Turin. The fourth professor of medicine, Giovanni Rizzetti, pursued a similar career, though at a rather lower level, occupying a series of positions in the administration of public health and education.[59] His career, though considerably less prominent than that of the other three professors, shows the same features of co-operation with the French regime, and notably with the regime of the Executive Commission which collapsed in 1802. It was this common background, as well as their ostensible grievances, that united the faculty in its attempts to discredit Balbo's authority.

The medical professors placed him and the French in a difficult position, for strictly speaking, Balbo was indeed in the wrong. But the French, on the other hand, fully intended that their politics of the restoration of the conservative nobility should be put into practice, and the revolutionary period forgotten. The deluded radicals of the 1790s were to be treated with consideration, but only so long as they consented to be relegated to their professional avocations. Special envoys to Piedmont had emphasized the point ever since annexation had been decided.

> Les savants et les hommes de lettres ont été pour la plupart très prononcés en faveur de la révolution français et . . . la réunion à la France, et la liberalité de leurs idées a dû les rendre odieux au parti royaliste. Ils m'ont paru aujourd'hui rentrés dans le cercle d'où ils n'auroient pas dû sortir, et diriger toute leur exaltation vers les sciences et les arts. Il est donc bien essentiel que le gouvernement leur accorde une éclatante protection, et sourtout qu'on ne leur oppose pas pour les éloigner des places, la hardiesse de quelques principes philosophiques indiscrètement avoués pendant l'effervescence de la révolution. J'ai recommandé cet esprit de tolérance aux autorités locales, persuadé que l'instruction publique seroit bientôt dégradée par la mediocrité si l'on a voulu juger trop rigoureusement des véritables savants que le malheur du temps ou

---

59 For Buniva and Balbis, see 'Studi Pinerolesi', *Biblioteca della Società Storica Subalpina*, ed. B. Vesme *et al.* I (1899), 305–77; Bianchi, *Storia* . . . III, 94; F. 17. 1609, Ecole de Médecine, Personnel, Etats de Service. For Rizzetti, ibid. letter of 20 Aug. 1809.

l'erreur ont quelquefois entrainés au delà des bornes de la moralité et de la sagesse.[60]

But the radicals refused to accept this measure of recognition, and agitation in the medical faculty in Turin eventually became so violent as to be viewed by the French as a possible cause of political upheaval. Fourcroy was warned, 'Dans un pays comme celui-là une étincelle peut causer une incendie; une discussion classique peut faire naître des partis.'[61] Tension increased after Bonvicino made the whole affair public by venturing into print with a pamphlet sharply attacking Balbo. Menou warned Bonvicino that such actions endangered his attempts to 'Bannir cette diversité de pouvoirs et d'autorité qui est le plus souvent mère à l'anarchie, dont nous avons fait une si terrible épreuve'. In the same month, June 1807, with Menou presiding *ex officio*, the governing body of the university formally condemned Bonvicino and his party. Nevertheless, agitation continued. Balbo tried, as he had done in the case of Balbis, to moderate the conflict by the use of his powers of appointment. In 1809 he recommended that the successor to one of the chairs in the faculty which had fallen vacant should be chosen not by competition, as the regulations demanded, but from among names which he himself would submit for the Grand-Master's approval, 'pour neutraliser autant que possible les effets de l'esprit de parti'. As a final manoeuvre, Balbo and Menou used the pretext of Bonvicino's election to the *Corps Législatif* in 1812 to oust him from his position as dean of the faculty, and replace him by the more docile professor Bellardi. Bonvicino's protests went unheeded, since the Imperial University had by then decided, with whatever misgivings, to throw the weight of its support behind Balbo, Menou, and the conservatives.[62]

Once Menou had given his support to Balbo, there was in fact little else that could have been done without undermining the authority of both men, and hence

---

60 AF. IV 1025, 'Mission du Citoyen Laumond dans la 27e. Division Militaire . . . 27 frimaire an XI, p. 17, 'Esprit Publique'. For Laumond, see J. Godechot, *Les Commissaires aux Armées sous la Directoire: Contribution à l'Etude des Rapports entre les Pouvoirs civils et militaires* (2 vols. Paris, 1937), II, 344.

61 F. 17. 1609, Ecole de Médecine, Sédillez to Fourcroy, 8 Sept. 1807.

62 Bonvicino's pamphlet is in F. 17. 1609, Organisation, Affaire de M. Bonvoisin (*sic*) from which the preceding paragraph is also drawn. Balbo's rejoinder of 1807 is in the Wellcome Museum of the History of Medicine, London, MS. 1040. The prominence of doctors in the politics of the revolutionary period has yet to be satisfactorily explained. The great advances in medical science in this period, in the work of such pioneers as Xavier Bichat, may have encouraged an awareness of new ideas and an openness to change, particularly changes emanating from France. More specifically, a strong connection is observable between the theories of perception which underlay the new medicine, and the high value placed on individual interpretation of the visible world, unconstrained by tradition and religion. See S. Moravia, 'Philosophie et médecine en France à la fin du XVIIIe. siècle', *Studies in Voltaire and the eighteenth century*, LXXXIX (1972), 1089–1151.

of the conservative group. Balbo's own frequent threats of resignation if he were not supported by Paris would in any case have left the central authority in Paris with little freedom of choice. It was a great disadvantage to the policy of installing conservative native officials that, once appointed, they were difficult to dispense with or dispose of, for the very reasons which had dictated their appointment in the first place. In this respect, the French were captives of their appointees, even if these appointees, like Balbo, devoted much of their time to the obstruction of the French system. In the case of the Imperial University, the conservatism of its staff ensured that no more flexible compromise with the remaining republicans was possible. The French remained trapped by the paradoxes of their own political solution to the problems of Piedmontese civil administration within the Empire.

These paradoxes did not vanish with the collapse of the Empire; in fact they became increasingly important during the succeeding decade. A substantial part had been played in revolutionary politics in Piedmont by men engaged in academic life. The new importance of this occupational group in political life has only recently been realized, and never fully explored. This is partly due to the surprising lack of biographical analysis of the protagonists of the revolutionary period. It is also only recently that the Napoleonic period in Italy has ceased to be dismissed as a sterile interlude between revolution and Risorgimento; little attempt has been made to establish continuities between all three periods. One such linking thread is provided by the gradual rise of professional men and university teachers as a political force to be reckoned with. Between 1814 and 1830, great changes took place in the organisation of the institutions of learning and science and their relation to the state. At the same time, it became possible as never before to diffuse facts and opinions fast and effectively to most classes of society. The control of the diffusion of knowledge thus became a newly acute political problem, and educational institutions became new political battlegrounds. In Italy, those who were involved with these problems, as Balbo was during the upheaval in Turin in 1821,[63] faced a political situation moulded by the actions of the previous regime.[64]

The early Risorgimento held many contradictory ideas on the subject of the Napoleonic regime. The Empire had brought with it a high degree of the very

---

63 P. Egidi, *I. Moti . . .*; Lenore O'Boyle, 'The problem of an excess of educated men in Western Europe, 1800–1850', *Journal of Modern History,* IV (1970), 471–95.

64 Most of the appointments made in the Universities by Napoleon were maintained by the restored monarchies. This surprising degree of continuity also helped to keep in being the attitudes towards education and politics generated by experience of the Imperial regime. For the careers of Benjamino Sproni at Pisa and Daniele Berlinghieri at Siena, see A. Zobi, *Storia Civile della Toscana dal MDCCXXXVII al MDCCCXLVIII* (5 vols. Florence, 1850–2), vol. IV and Appendix. For Gerolamo Serra at Genoa, see R. Boudard, *L'organisation de l'Université . . . dans l'Académie Impériale de Gênes entre 1805 et 1814* (Paris–The Hague, 1962). Appointments in the former Dutch departments of the Empire showed the same degree of continuity. See S. Schama, 'Schools and Politics in the Netherlands, 1796–1814', *Historical Journal,* XIII (1970), 589–610.

territorial unification and administrative rationalisation demanded by liberals in the 1820s. Former employees of the French were to the fore in many of the revolts of 1821.[65] Yet at the same time the subordination of the Italian people within the Empire was repugnant to aspirations of national unity and independence. The Empire had also established new lines of political cleavage. Conservative collaborators had had to turn to the French to restrain the university radicals, as Balbo had turned for help against Bonvicino. In the period of the Enlightenment, things had been very different. The Italian ruling houses and the intelligentsia had worked together for reform.[66] In the 1790s the position had started to change. Fearful that events in revolutionary France would spread to their own territories, the monarchs drew back from their support of reform and, in doing so, isolated many of the intellectuals who had been their strongest supporters. When the French began to invade Italy, it was thus not surprising that this group should provide much of the impetus behind the revolutions which paved the way for the final collapse of many of the old monarchies. But the French were not grateful allies. The radicals and republicans found themselves exiled or relegated to the obscurity of the classroom. When Napoleon reinforced the position of the conservative nobility in administration, he did nothing to quieten the hostility of the moderates, and in fact could have found no better way of provoking such outbursts of hostility as that between Balbo and the medical professors. No real policy of *amalgame* was put into effect. The disaffection of the republican intelligentsia within the University from the ruling authority was thus maintained, and carried over without difficulty after the restoration of the ruling houses once so gladly served by the universities.

It was in the light of this development that conservatives and liberals were to face each other in the 1820s. The demands of conservative collaborators strongly affected the functioning of the Napoleonic Empire; the way in which they were met influenced many of the characteristics of the politics of the Restoration period in Italy.

---

65 E.g. G. T. Romani, *The Neapolitan Revolution of 1820–1821* (Evanston, 1950).
66 N. Carranza, 'L'Università di Pisa e la formazione culturale del ceto dirigente toscano del sette-cento', *Bollettino storico pisano,* xxxiii–xxxv (1964–66), 469–537; Cochrane, *Tradition and Enlightenment. . . .*

2

THE LANGUAGE OF
NATURAL POWER

The "éloges" of Georges Cuvier and the public language of nineteenth century science

From: *History of Science*, 16 (1978), 153–178. © Sage Publications.

In this paper I examine the *éloges* of Georges Cuvier not simply as a collection of exemplars of the figure of the ideal natural philosopher, but as an arena in which different explanations of the kind of power at the disposal of the natural philosopher untidily contend. The closely related literary forms of eulogy, panegyric and hagiography are all concerned with the chemistry of certain forms of moral authority.[1] Eulogies of natural philosophers face a problem of special complexity and interest because they attempt to describe how authority is gained, not only from the ratification and consent of human groups, but also from the encounter of the human individual with the natural world. On one level, this encounter put the natural philosopher in a peculiarly intimate contact with a universe whose organization was regarded as bearing a close analogical relationship with its divine, beneficent and omniscient Creator, a Creator who was also responsible for the ordering of human society. To some extent, the encounter with nature could thus be regarded as a form of encounter with the holy. But the encounter with nature could also be a shocking one. In the early nineteenth century, easy analogies between the natural world and the divine had begun to break down. The universe seemed less an ordered system set in motion by a benevolent power, and more a dynamic play of possibly autonomous forces.[2] This meant that the position of the interpreter of nature, the natural philosopher, similarly became problematic. Where did his authority derive from, if nature could not be strongly identified with an aspect of the holy? What means therefore had he to channel and rationalize the emotions unlocked by contact with the beauty, power, and complexity of nature, and translate them in ways acceptable to the various human groups to which he belonged?[3]

The academic funeral *éloge* in these circumstances operated as a facilitating device. Its highly formalized literary structure provided strong outer walls to contain the conflicts and uncertainties of the groups and individuals to whom it was addressed. *Eloges*, firstly, are only produced on the occasion of a death.

DOI: 10.4324/9781003038085-3

Their function as *rites de passage* allowed preoccupations to objectify around the incomprehensibility of death. Secondly, phenomena experienced as personally, socially and politically confusing by their audience could be rendered explicable by establishing a connection between that audience and ideal activities outside any social production, and in particular outside any contradiction. This was achieved by the way in which the idealized, semi-pastoral language of the *éloges* allowed fact and symbol to flow into each other without actual explanation and definition. *Eloges* could therefore deal with the issue of power, while seeming to, and because seeming to, celebrate an image of the natural philosopher as a-political and a-social. In an age when the nature of political power had changed radically, violently, and rapidly, it is not surprising that Cuvier's *éloges* should be preoccupied by the distinction between legitimate and illegitimate forms of power, and by the relationship between the authority gained by the natural philosopher in his encounter with nature, and the other forms of power exercized in human groups. This interest was heightened by the emergence in the revolutionary period of the intelligentsia as a new and major political force. The question of how intellectual authority and political power, *gloire* and *pouvoir*, were linked was thus acutely important to Cuvier's audience both in terms of their personal experience, and of the political options open to them.

It is in the light of these considerations that I shall discuss the *éloges* written by Cuvier for the Institut and the Académie Française.[4] Such notices were read aloud at the annual public sessions of the learned bodies concerned. These occasions attracted the attendance of most of the political and intellectual *élite* of Paris, an audience whose composition reflected the importance of the Institut as a means of cultural propaganda in the Napoleonic state. Even without the changes which the revolutionary period had brought to the public role of the scientific community, the writing of the *éloges* would thus have had to have been undertaken as a calculated public gesture. The nicety of the calculation rested on the fact that the *éloges* discussed issues of various kinds of power in a public context, yet did so by means of a structure of images which professed to endorse a decisively a-political, solipsistic image of the ideal natural philosopher. This is the central tension in the *éloges* as literary and public creations, and it is this tension which makes the use of the *éloges* as quarries for unproblematic, coherent images of the ideal natural philosopher a somewhat naive undertaking.[5]

How does Cuvier set about working through these tensions? A fairly detailed examination of the language of the *éloges* is perhaps the best way of approaching this question. Cuvier approaches the issue of what is legitimate, 'clean', authority, and how it may be gained by the right kind of contact with nature, through the use of the evaluation of character as a basic normative tool in the delineation of an acceptable style of scientific enquiry.[6] This remains true, whatever the surface contradictions in the various *types* of character portrayed as belonging to the ideal natural philosopher. Built into this ideal character was a strong bias in favour of naive observation. Blessed are the pure in eye.[7]

This 'clean' power, which both gives, and results from, direct access to the secrets of nature, is attained by a life aloof from the distraction, corruption and intrigue of courts and polite society, a life "completely solitary, completely scholarly", "unmindful of glory, ignorant of the world".[8] This stereotype emphasizes as a basic concern the apartness of the natural philosopher from human society and its political goals of the struggle for fame and fortune. It does so by the recurrent use of words such as "glory", "the world", "intrigue" and "fame", in constant opposition to words such as "solitude" and "retreat". This structure of images establishes close links between the *éloges* and pastoral.[9] This is a literary structure which goes out of its way to establish its independence from the structures of dominance and dependence which organize human groups. Because the natural philosopher is seen as using only the 'clean', non-exploitative power derived from contact with nature, he cannot use his special authority to challenge other forms of power, and, equally, is removed from the possibility of suffering the direct exercise of 'dirty' political power.

But this is only half the story. It would be ironic indeed if the period which saw the emergence of the natural philosopher as a political figure were characterized in its celebrations of the dead by its endorsement of an a-political ideal of science. Cuvier was indeed struggling with the deep tensions of the *éloge* as a public literary form at a time of rapid political change, and consequently in his funeral speeches the 'pastoral' description of the life of the ideal natural philosopher has to struggle with an explanation of the motivation of scientific activity in which the drive for power, fame, and recognition in the arena of public estimation is awarded a very large place. Many of his subjects were born in humble circumstances or even in poverty, and Cuvier often points out how the fulfilment of a scientific vocation necessarily entailed the ruthless and hectic overcoming of massive social disadvantages. The calm satisfactions of the retiring scholar have little to do with this struggle to fit a grid of public measurements of ambition and success. At many points, Cuvier sees the typical natural philosopher "grappling not only with nature but with fortune, and by dint of perseverance winning permanent victories over both of them".[10] In this view the "ardour of the will"[11] counts for everything. It is the least pastoral of virtues.

Cuvier also views this struggle for recognition by humble talent in an aristocratic society as the major factor precipitating many men of science onto the turbulent waters of revolutionary politics. Lack of public recognition, leading to a justifiable sense of hardship, impelled Antoine Fourcroy, for example, into participation in extreme radical movements in the 1790s. Cuvier's analysis of Fourcroy's career in fact brings us close to the heart of the conflict in the *éloges* between the kind of power derived from contact with nature, and that derived from the exercise of power in the human group. For Cuvier, the moral of Fourcroy's life was the "power of hard work and the will to dominate fortune, just as much as the inability of fortune to confer happiness".[12] Here, the "fortune" which it had been the duty of the aspiring natural philosopher to master, becomes in a political context a delusory and harmful goal for a natural philosopher to pursue. Cuvier implies that scientific success cannot be obtained in adverse social or political circumstances except by the exercise of the very qualities which bring unhappiness and disillusion in the political arena.[13]

The same man might have to use both 'clean' and 'dirty' power in his life. Thus it is not surprising that in many of Cuvier's *éloges* key words such as "*gloire*" and "*fortune*" have ambiguous, shifting connotations. Cuvier frequently tries to disguise their links with 'dirty' power by using changes of context within single sentences to imply changes of meaning. This is the point of his frequent references to the long roll of past Academicians "who gained glory by enlightening the world, unlike other men, who sought to gain glory by devastating it".[14] The natural philosopher, in other words, is distinguished by using his authority or "*gloire*" in a 'clean' way. But this is different again from the kind of "glory" which Cuvier posited as the necessary spur to the endeavours of the man of science: the reward of the conquest of social obstacles to the fulfilment of a scientific vocation by an almost Machiavellian "ardour of the will".

Another key word in the establishment of the pastoral view of the natural philosopher is "*solitude*". Solitude defines the scenario in which the natural philosopher lives "in his retreat", as the stock phrase goes. The setting is always a rural one; Bosc, for example, lived "far from plots and intrigues, a life amongst the woods which he had deeply loved from his earliest youth".[15] This pastoral solitude is often associated with the virtues of probity and dignity also ascribed in pastoral to the country-dweller. Cuvier describes the agronomist Cels, for example, as "a man of the fields, a stranger to the ways of society, always inflexible in his support of the true and the just".[16] Solitude also had the power to cushion the natural philosopher against social isolation, "when the unspeakable charm of the search for truth makes one despise poverty and neglect".[17] Solitude in fact is an essential part of the chemistry of the moral authority of the natural philosopher, because it decisively established his remoteness from the ordinary pressures of human groups and allowed a 'pure' and immediate reaction to the natural world, a reaction in turn endorsed by contemporary theories of how good 'naive' observation of the natural world takes place. Reliance on pastoral images and presuppositions also enabled Cuvier to make deeper-level generalizations about what it is that confers status in a human group preoccupied with the problem of the translation of the encounter with the natural world into terms acceptable both to themselves and to a wider outside audience.[18]

Yet in the last analysis, the meaning of "*solitude*" in the *éloges* is no less ambiguous than that of "*gloire*" or "*fortune*". Taken to extremes, solitude and retreat undermine the natural philosopher's ability to produce a valid translation of the encounter with nature. Cuvier's *éloge* of the botanist Michel Adanson, for example, ascribes many of the weaknesses of his theory of classification to his isolation from the rest of the scientific community. The same theme dominates the *éloge* of Lamarck. Lack of human contact led to fatal flaws in his work of which he was himself unaware. "In this respect he resembled so many other solitary men, who never seem to have doubted themselves, because they had never had the opportunity of being contradicted." In a striking image, Cuvier went on to assert that excessive solitude led Lamarck to form metaphysical systems, "like the enchanted palaces of old fairy tales, which disappear when their talisman is broken". In attacking Lamarck on these grounds, Cuvier was also attacking the association of solitude with the figure of the 'martyr to science' extensively used in the historiography of Lamarck. In Cuvier's hands, Lamarck's solitude became an indictment instead of an endorsement. Friendship is "the only enjoyment which our noble élite of humanity can never renounce", not even for the "uncertain future of glory itself".[19]

The use made by Cuvier of the key words in the pastoral image of the natural philosopher thus strongly reflects the tensions within the *éloge* form, while their ambiguity enabled whole series of concerns about the conduct of natural philosophy to be marshalled under economical forms of reference. It is also clear that Cuvier deliberately exploited the shock-value of overtly political statements in the pastoral context of the *éloges*. His friend the German mathematician Christian Pfaff has left an eyewitness account of one such occasion during the public reading of Cuvier's *éloge* of the agronomist Gilbert in 1801. Pfaff recalled:

> I shall never forget the effect which this piece produced on a numerous and brilliant audience, when they realised that in praising Gilbert, Cuvier was attacking the new Sejanus of our times. Especially deafening applause greeted his statement that "already the men who worship power hardly make the barest offerings of respect to genius, so much does the power which shapes thought appear to them inferior to the power which dispenses fame and fortune".[20]

The exploitation of such tensions and paradoxes in the *éloge* form had also been given a new twist by the way in which pastoral images of the scientific life had themselves entered an explicitly political context in the revolutionary period. The work of writers such as Bernadin de St Pierre and Brissot de Warville had associated these very pastoral images with the attack on the old Académie des Sciences, many of whose former members were present in Cuvier's audience at the Institut. In his novel *La chaumière indienne*, Bernadin de St Pierre endorsed the idea that knowledge of nature was not the prerogative of specialists, such as members of the Académie des Sciences, but descended on all those who lived pure, and preferably rural lives. Ideas like this early linking of the 'natural' and the personally 'authentic' were powerful in fuelling the movement which culminated in the closure of the Académie des Sciences in 1793 as an élitist, anti-democratic, and hence anti-natural corporation.[21] Thus by 1800 the pastoral elements in the ideal of the natural scientist had taken on some rather menacing connotations for Cuvier's audience. This is the point of the otherwise gratuitous aside in his *éloge* of the chemist Darcet of 1802. He expatiated on the pastoral mediocrity of Darcet's life

> . . . never diverted from his labours in obscurity by the temptations of ambition, and who, devoted to the public good, never asked for any other reward but that of personal satisfaction; is not such a man a model in these times when the social fabric must be based on moderation and all the virtues of peace?[22]

In this passage, Cuvier turned the pastoral image away from the dangerous radical associations which it had acquired in the 1790s. The pastoral image once again reassured, because the ideal it presented neither threatened nor reproduced the patterns of dominance and submission present in human political society. In 1802, the establishment of the life Consulate for Napoleon had consolidated the political revolution of the past decade. A new era of hard, threatening militaristic government had opened, in which the individual was deprived of all the corporate, regionalistic buffers which had shielded him from the attentions of the state under the Ancien Régime. It was thus hardly surprising that an increasing need should be felt to define the realm of the scientist as at once inviolable and value-free, because operating in a different realm from that of human political activity.[23] Large-scale political change in the early nineteenth century was one of the major factors in the definition of the dominant ideology of science, as both inviolable and value-free.

Cuvier's awareness of the scars left by the Revolution on the organizing images of the *éloge* also extended to a conscious manipulation of the *éloge* to celebrate his own inclusion into the academic community. In the following passage, describing the reformation of the scientific community in 1795, on the foundation of the Institut on the ruins of the Académie des Sciences, he plays on the more emotive aspects of the ideal of the scientific community, against a political background which remains implicit in the experience of his audience. For the new-comer, he wrote, the first session of the Institut left an unforgettable impression:

> . . . what tears of joy, what mutual and pressing enquiries on each other's misfortunes and exiles, each other's occupations; what unhappy reminiscences of so many of their colleagues who had fallen victim to the executioner; what sweet emotion for those who, still young, and sitting for the first time at the side of men whose genius they had long respected learnt by this moving scene to know their hearts![24]

Cuvier, by describing his audience to themselves, had also rectified his relationship with them. A relative new-comer to the scientific community, and the youngest member of the Institut apart from Napoleon himself, he may well have felt the need to justify his position of authority by thus identifying himself with the old Academy of Sciences. In doing so, he also established his right to manipulate the political overtones which the pastoral images of the *éloges* had acquired by 1795.

In turn, these overtones mirrored other movements of thought in the eighteenth century, all of which were concerned with the problem of how the exercise of right reason was related to certain ideals of character. The growth of the idea of 'the genius' in writers such as Diderot had included the view that intellectual achievement of any kind is typically accomplished by exceptional individuals living outside social bounds. This is obviously related to the ideal of pastoral isolation which was supposed to define both the life of the true natural philosopher and the quality of his perception of nature. Diderot's work also helped to reinforce the idea that natural philosophy was different from other professions, because it implied that genius was inborn and hence that the special authority conferred by knowledge of nature could not be acquired by a conscious process of learning.[25] Behind this again lay a whole theory of how a relationship with nature is built up, by brilliant sympathetic insights, rather than by the accumulation of 'facts', which could be performed by any well-trained plodder.[26] But such ideas also challenged the assumption that the Republic of Letters provided the ultimate court of judgment for the translation of the natural philosopher's perception of nature. Cuvier's account of the rôle of solitude in the life of the natural philosopher mirrors these ambiguities very closely. A final twist was provided by the fact that the idea of the Republic of Letters itself had had a very strong democratic significance ever since d'Alembert's essay of 1743 had stressed the necessity for the survival of the intellectual classes of the abolition of hierarchies based on birth or wealth.[27]

Thus in manipulating the seemingly simple and stylized ideal of the retiring natural philosopher, Cuvier was in fact entering upon a minefield of political and ideological associations, a minefield quite as explosive as the political changes in society at large which had changed the whole public context of the exercise of intellectual authority. An understanding of the language and structure of the *éloges* can be seen as one way of approaching an analysis of the evolving public language of science. Through pastoral, science could be presented as neutral and therefore unchallengeable. Yet the ambiguities of the *éloge* also enabled Cuvier simultaneously to call up and to paper over the very anxieties about the generation and use of power which lay behind the drive to establish the value-freeness of science. Pastoral images and their derivatives, such as the image of the natural philosopher in the period under discussion, can be viewed as fictions or analogies; the lives of actual natural philosophers do not of course conform entirely to pastoral stereotypes, but it is possible to objectify many central concerns about the social, religious, personal and political demands of intimate contact with nature by reasoning 'as if' they did. In the later nineteenth century, both the value-freeness of science and the operation of intrusive centralized government were better established and more taken for granted. Correspondingly, the texture of obituary notices no longer included a fine thread of ambiguity in the discussion of the scientist's relation to the exercise of authority in the public arena. The structure of public expectations of obituary notices altered correspondingly, to screen out the relics of out-moded preoccupations. This explains why the numerous nineteenth century reissues of Cuvier's *éloges* most often reprint those studies where Cuvier permits himself asides of a political nature, or comments harshly on the character of the departed – studies such as those of Fourcroy and Adanson – and yet can in their editorial material still present Cuvier's work as respectable uncontroversial contributions to the history of science.

Modern difficulties with the *éloges* mirror this later nineteenth century incomprehension. As biographical sources they appear opaque, thin in factual content, and irritatingly bland and repetitive in form. The very fact of their existence in large numbers, and their evident acceptability for contemporaries increases our incomprehension. But *éloges* cannot really be understood in terms of a substitution for 'proper' biography. They existed, much more, to facilitate the passage of the groups to whom they were addressed from doubt to resolution. Public and explicit expression of private difficulties over the issues treated in *éloges* thus became less necessary, and the disruptive potentialities of public debate were thus contained. Elites were small and interconnected and could still maintain elaborate common languages such as those employed in the *éloges*.[28] This should make us beware of levelling accusations of hypocrisy or undue reticence at a community whose public language could still afford a large reliance on the implied and the unspoken. Overt discussions of tensions between image and reality only became either possible or necessary when pastoral was jettisoned as a credible mode of public utterance; and this in turn depended to a large degree on the increasing heterogeneity of intellectual and political élites as the century progressed.

We should also therefore not accuse Cuvier in particular of hypocrisy because he left no public statement on the tension between politics and science which dominated both his own life, and contemporary views of him. The best comment on these issues is to be found in the tensions and ambiguities of his own *éloges*.[29]

Above all, *éloges* are about death. They exist to help the living over the gap between their memory of the human imperfections of the departed, and the terrifying incomprehensibility of his departure. In the construction of a healing distance from death, pastoral found its hold as a medium for the discussion of the intimate, exciting, and yet often threatening encounters which formed the basis of natural, human, and supernatural authority, at their common focus in natural philosophy.

Acknowledgements

Those who contributed to the growth of the ideas contained in this paper include members of the Leicester Society for the History of Science and Technology, and of the Wellcome Unit for the History of Medicine, Cambridge, and in particular

W. H. Brock, Karl Figlio, Nicholas Jardine, Ludmilla Jordanova and A. G. Keller. Jonathan Mandelbaum provided information on matters relating to the Société Philomatique, and Jessie Sweet answered many enquiries on the papers of Robert Jameson. Martin Rudwick discussed with me problems relating to the diffusion of the *éloges*. I would also like to thank Peter Brown, Peter Dronke, Adrian Lyttelton, Partha Mitter, Roy Porter, Quentin Skinner and Richard Tuck; the staff of the Institut de France, for their guidance in the labyrinths of the Fonds Cuvier; and the University of Reading, during the tenure of whose Research Fellowship material for this paper was collected.

Notes

1 Here I should acknowledge my indebtedness to S. G. McCormack, "Latin prose panegyrics", *Revue des études augustiniennes,* xxii (1976), 29–77; and P. R. L. Brown, *The making of Late Antiquity* (to appear, Harvard University Press).
2 For Cuvier's own Hobbesian views, which contained an implicit attack on the idea of a beneficent divine order reflected in the human and the natural world, see *Eloges,* i, 91–92: ". . . nous retrouvons parmi eux [the animals] le même spectacle que dans le monde; quoi qu'en aient dit nos moralistes, ils ne sont guère moins méchants ni guère moins malheureux que nous; l'arrogance des forts, la bassesse des faibles, la vile rapacité, de courts plaisirs achetés par de grands efforts, la mort amenée par de longues douleurs, voilà ce qui règne chez les animaux comme parmi les hommes." For the part played by Cuvier in establishing the autonomy of the natural world, see M. Foucault, *Les mots et les choses* (Paris, 1966), 275–92.
3 The problem of translation was intensified by the existence in the late eighteenth and nineteenth centuries of two different styles of response to the encounter with the natural world. The aesthetic of the sublime endorsed the primacy of an individual, untranslatable encounter with nature. On the other hand, ideals of rationality and public utility as justifications for the pursuit of natural philosophy, are more readily acceptable in public statements of the rationale of science, and are accordingly stressed in the *éloges*. For another example of contemporary attempts to reconcile these two strands of thought, see the description of Humphry Davy in D. M. Knight, "The scientist as sage", *Studies in Romanticism,* vi (1967), 65–88.
4 Before Cuvier's arrival in Paris in 1795, he had taken a keen interest in the collection of data on the lives of eminent contemporary natural philosophers. In 1795, he joined the Société Philomatique in Paris, and produced several funeral *éloges* on its members. It is difficult to say whether this was an important factor in his appointment in 1803 as one of the Perpetual Secretaries of the Institut; however, it is certain that his duties in that post involved the composition of a publicly read obituary notice on all those of his colleagues who had died in the preceding half-year. After his election in August 1818 to membership of the Académie Française, he performed similar duties for that body. Thus Cuvier became the main agent of biographical evaluation, not only for the two major bodies of official French culture, but also for one of the most important Parisian private scientific societies. He also reached a wider audience, for between 1811 and 1828, he contributed forty-two articles on men of science to the first edition of the *Biographie universelle*. The *éloges* also achieved a wide diffusion. They automatically appeared in the official organ of the Institut, the *Mémoirs,* and could also be reprinted in the intellectual journals such as the *Magasin encyclopédique*. Between 1819 and 1827 Cuvier published a collected edition of the *éloges* in three

volumes. They were reprinted twice in the 1860s under the editorship of Cuvier's
protégé Pierre Flourens. They appeared in edifying selections for children as late as
1911, and are still used in French schools as stylistic models. In the nineteenth century,
individual *éloges* were frequently used as introductions to works of popular natural
history, or to scientific biography. Robert Jameson's translation of fifteen *éloges* in
the *New Edinburgh Review* between 1821 and 1831 played an important part in their
diffusion in England. Sarah Lee's standard *Memoirs of Baron Cuvier* (London, 1833),
devotes nearly a hundred pages to précis and paraphase of the *éloges*. The publication
history of the collected editions and of individual *éloges* is traced in the bibliographi-
cal section of this paper.

5 For examples of such attempts, see O. Sonntag, "The motivations of the scientist: The
self-image of Albrecht von Haller", *Isis,* lxv (1974), 336–51; S. Delorme, "Des éloges
de Fontenelle et de la psychologie des savants", *Mélanges G. Jamati* (Paris, 1956).

6 See D. Outram, "Scientific biography and the case of Georges Cuvier", *History of science,* xiv (1976), 101–37.

7 See M. Foucault, *Naissance de la clinique* (Paris, 1963), 107–8. Hence the ease with
which modern indoor laboratory science has often been identified in popular writings
with 'bad' or 'dirty' science.

8 *Eloges,* ii, 311 (Werner); iii, 95 (Duhamel).

9 For other specific examples of the connection between pastoral and the tension between
otium and *negotium,* retreat and the world, contemplation and involvement, see J. Mat-
thews, *Western aristocracies and Imperial Court, AD 364–425* (Oxford, 1975), 9–12;
M. Sullivan, *The birth of landscape painting in China* (London, 1962), 1–24; S. G.
McCormack, *op. cit.* (ref. 1).

10 *Eloges,* ii, 269 (Tenon).

11 *Eloges,* ii, 380 (Riche).

12 *Eloges,* ii, 6 (Fourcroy). Here the question of Fourcroy's involvement in Lavoisier's
execution also undoubtedly cast a shadow.

13 This is a modification of the classic Stoic position of the retreat into *otium* and pastoral
mediocrity as a response to oppressive government, and as a means of gaining the calm
of spirit necessary to reflect upon that government. Unlike the Stoic, who used his calm
contemplation to organize his perception of unjust government, Cuvier's natural philosopher exists in a world whose rural, pastoral elements isolate him completely from
the problems of personal response to political problems. This emphasis on the inviola-
bility of the Stoic is also present in Rousseau's reworking of classic Stoic themes. See
K. F. Roche, *Rousseau: Stoic and romantic* (London, 1974). It also renders impossible
any recourse to references to a Cincinnatus-style passage from pastoral life to the exercise of 'clean' political power.

14 *Eloges,* i, 83.

15 *Académie des Sciences, Mémoires (Histoire),* x (1831), 204.

16 *Eloges,* i, 254.

17 *Eloges,* i, 169.

18 For the preoccupation of the medical community with similar problems of translation
posed by the 'interrogation' of nature, see M. Foucault, *op. cit.* (ref. 7).

19 The notorious reputation of Cuvier's *éloge* of Lamarck should not obscure the fact
that if anything it is his appreciation of Adanson which is the more critical. Images of
Cuvier and Lamarck have been made to interact in the historiography in ways which
imply an expectation of conflict between them, and which have become entwined with a
highly emotive debate on the origins of the theory of evolution. Cuvier's relations with
Adanson, on the other hand, have never received the systematic analysis demanded by
Cuvier's own acknowledgement in this *éloge* of his own debt to Adanson's theory of

classification. The *éloge* was originally published in *Mémoires de l'Académie Royale des Sciences,* xiii (1835) and the passages quoted are on pp. 1–11; that from the *éloge* of Adanson at *Eloges,* i, 304. For another example of the historiographical linkage between solitude and the 'martyr to science', see B. G. Beddall, "'Un naturalista original': Don Félix de Azara, 1746–1821", *Journal of the history of biology,* viii (1975), 15–66.

20 L. Marchant (trans.), *Georges Cuvier: Lettres à C. H. Pfaff sur l'histoire naturelle, la politique et la littérature, 1788–1792* (Paris, 1858), 33. Juxtaposition brings out sharply in this passage the two senses of power, one referring to 'clean' power and the other to 'dirty' power. It is unclear who Cuvier had in mind as the "new Sejanus"; it is difficult to assimilate the powerful but low-born favourite of the Emperor Tiberius entirely with any of the obvious candidates such as Cambacérès, Chaptal or Lucien Bonaparte.

21 C. C. Gillispie, "The Encyclopédie and the Jacobin philosophy of science", *Critical problems in the history of science,* ed. M. Clagett (Madison, 1962), 255–89. See also R. Sennett, *The fall of public man* (Cambridge, 1977) for a discussion of nineteenth century difficulties in the location of 'authentic', 'natural' response in the public arena.

22 *Eloges,* i, 85.

23 *Cf.* S. Diamond, *The reputation of the American business man* (Cambridge, Mass., 1955), 181, for a similar utilization of the conventions of obituary notices on individuals to prohibit criticism of the nature of areas of activity as such – in this case that of the operation of the American economic system.

24 *Eloges,* i, 305.

25 H. Diekmann, "Diderot's idea of genius", *Journal of the history of ideas,* ii (1941), 151–82.

26 Most of the historiography of Cuvier himself centres on this debate between the possibility of a pure, 'naive' vision of nature which emphasizes the collection of 'facts', and the more intuitive view, which laid great stress on conceptualization *before* the 'facts' were collected. See M. Foucault, *op. cit.* (ref. 7). It is against this background that 'Baconian' research programmes in the natural sciences in this period should be assessed.

27 Jean de la Ronde d'Alembert, "Essai sur la société des gens de lettres et des grands, sur la réputation, sur les Mécènes, et sur les récompenses littéraires", *Oeuvres complètes* (Paris, 1790), vi, 335–73.

28 For another example of a highly evolved common language maintained by a small-scale élite, see S. G. McCormack, *op. cit.* (ref. 1), 53: orators and panegyrists in fourth century Gaul "spoke by allusion, implication and symbol".

29 *Cf.* S. Diamond, *op. cit.* (ref. 23), 177, for the use of idealized figures "as a focus around which consensus may be attained, and as a stimulus for the voluntary acceptance of the discipline implied in the behaviour he symbolizes". Cuvier rarely if ever uses the term 'genius', and seems to draw little upon the tradition originating in Vasari, and continued through Diderot, of the genius as an individual of exceptional emotional force, thereby often posing danger to others. Genius-as-abstraction similarly plays little part in his view of the natural philosopher. See R. Wittkower and M. Wittkower, *Born under Saturn* (London, 1963), for the general history of this myth of genius in the artist especially. For Cuvier, the natural philosopher is above all an interpreter, one whose vision interrogates nature and unveils her secrets. For his rejection of contemporary Chateaubriand-inspired ideas of the wild-eyed poet as the true interpreter of nature, see his article in *Le moniteur,* 3 novembre 1807, p. 1186. Here, Cuvier set out his own more restrained and classic view of the community of natural philosophers, who "mettent

leur gloire à dévoiler quelque partie des secrets de la nature; comme ils sont calmes dans ce sentiment intime qu'avec du travail ils ajouteront toujours quelquechose à cet édifice imposant de la science que les siècles élevent; dont toute l'humanité profite et dont rien n'ébranle la partie une fois fondée sur les faits. Ils sentent qu'ils n'ont des juges que leurs pairs répandus dans tout ce monde civilisé et qu'aucun prestige ne pourrait ni éblouir ni gagner. Ils ne s'exposent donc ni à être trompés eux-mêmes par des applaudissements aveugles ni à faire des démarches honteuses pour s'en procurer de factices".

Bibliography of the biographical writing of Cuvier

Abbreviations

AMNHN: *Annales du Muséum national d'Histoire naturelle.*
BIFC: Bibliothèque de l'Institut de France, Fonds Cuvier.
BU: *Biographie universelle*, ed. Michaud (52 vols, Paris, 1811–28).
EH: *Eloges historiques* (3 vols, Paris/Strasbourg, 1819–27).
MAS: *Mémoires de l'Académie royale des Sciences.*
MI: *Mémoires de l'Institut de France.*
MMNHN: *Mémoires du Muséum national d'Histoire naturelle.*
NEPJ: *New Edinburgh philosophical journal.*

Contents

A. Collected editions.
B. Individual studies in chronological order of writing.
C. Doubtful *éloges*.
D. Alphabetical index of biographical subjects.

A. Collected editions

Georges Cuvier, *Recueil des éloges historiques lus dans les séances publiques de l'Institut royal de France, précédé des reflexions sur la marche actuel des sciences* (3 vols, Paris/Strasbourg, 1819–27).
The fourth volume of this edition mentioned in Sarah Lee, *Memoirs of Baron Cuvier* (London, 1833), 214, was never printed. BIFC 3303/59, 25 April 1818, puts the maximum number to be printed of this edition at 2000 copies for vols i and ii. Information on the printing of the third volume in 1827 is lacking.
Georges Cuvier, *Eloges historiques précédés de l'éloge de l'auteur par M. Flourens* (1 vol., Paris, 1860: Bibliothèque classique des célébrités contemporains).
This edition includes the studies of Daubenton, Priestley, Adanson, Fourcroy, Cavendish, Pallas, Parmentier, Rumford, Haüy, Berthollet, and Davy. The text is in all cases identical with that of the 1819–27 edition.

Georges Cuvier, *Recueil des éloges historiques lus dans les séances publiques de l'Institut de France. Nouvelle édition précédée de l'éloge historique de Georges Cuvier par Flourens* (3 vols, Paris, 1861).
This complete edition was reprinted in 1867; the text is again unaltered from that of 1819–27.
Nouvelle bibliothèque populaire: les grands savants français: Cuvier (Paris, 1891). Text again unchanged.
Contains the studies of Daubenton and Adanson.
Léon Chauvin, *Georges Cuvier: Savants français: éloges historiques prononcés à l'Institut de France, précédés d'une notice par Léon Chauvin, ancien inspecteur primaire, directeur honoraire d'école normale* (Limoges, n.d. [? 1910]).
Contains the studies of Daubenton, Gilbert, Darcet, L'Héritier, Fourcroy, Parmentier, Tenon, Haüy, Vauquelin. Text unchanged.

B. Individual studies in chronological order of writing

1. RICHE DE PRONY, Claude-Antoine-Gaspard (1762–5 September 1797).

   EH, ii, 377–424.
   "Eloge du citoyen Riche, lu à la séance générale de la Société Philomatique de Paris, le 23 frimaire an VI [13 December 1797]", *Rapports généraux des travaux de la Société*, vol. ii, 1793–.
   Also separately printed, paginated 169–228, n.d.
   Cuvier based his account on Riche's journals. BIFC 3221/4 contains letters of acknowledgement from his family. The article in BU, vol. xxxviii (1824), by his brother, is heavily reliant on EH.

2. BRUGIERES, Jean-Guillaume (1750–1 October 1799(?)).

   EH, ii, 425–42; an extract from "Notice biographique sur Brugières lue à la Société Philomatique dans la séance générale du 30 nivôse an VII [19 January 1799]", also in *Société Philomatique de Paris, Rapports généraux des travaux de la Société*, vol. iii, 1793–.
   BU, vi (1812), 91–92.
   Cuvier does not refer to the study in EH in his BU article. The speech to the *Société Philomatique* is the only source for Brugières's life which asserts that he was a member of the *Société*. Cuvier's BU article is silent on this point, which is not elucidated in Michaud's second edition. Biographical sources also differ considerably on the date of Brugières's death. Cuvier's BU article gives 1 October 1799, but his study in EH mentions 11 vendémiaire an VII (2 October 1798). The biographical index of the Académie des Sciences gives the date as 10 vendémiaire an VII (1 October, 1798). Unfortunately, Cuvier's oration to the *Société Philomatique* was not written into its minutes.

3. DAUBENTON, Louis-Jean-Marie (1716–31 December 1799).

    EH, i, 37–80, read 5 April 1800. Material relating to the composition of this *éloge* is to be found at BIFC 3174, 3175, 3176. Cuvier mentions the reading in his "autobiography" as the payment of a debt towards the older generation of natural philosophers in France. See MS Flourens 2598(3) of the Library of the Institut de France.

    The text of EH is printed in MI, iii (Hist.), 1800, 69–100; separately printed as "Notice sur Daubenton lue à la séance publique de l'Institut national de France du 15 germinal an VIII" (Paris, an IX, 32 pp.).

    Also reprinted in *Oeuvres complètes de Buffon* (44 vols, Paris, 1824–32); and in *Oeuvres complètes de Buffon* (7 vols, Paris, 1737–9; reprinted 1848, 1853, 1858).

    Cuvier's *éloge* of Daubenton also appears in all the collected editions cited in Section 1.

    It was also translated into English by Robert Jameson, NEPJ (April–September 1828), 1–22.

    BU, x (1813), 569–71.

4. LEMONNIER, Louis-Guillaume (1717–7 September 1799).

    EH, i, 83–109, read 7 October 1800.

    Same text in MI, ii (Hist.), 1800, 101–17.

    Separately printed as "Notice historique sur L. G. Lemonnier, lue à la séance publique de l'Institut national de France, du 15 vendémiaire an IX" (Paris, an IX, 18 pp.).

    Materials used by Cuvier in the writing of this *éloge* are to be found at BIFC 3177, 3178; some have also been reprinted in [H. Dehérain], *Journal des savants* (December 1906), 625–32, including a memoir of Daubenton by André Michaux dated 22 brumaire an VIII (13 November 1799). Cuvier shows considerable disrespect for the argument from design on p. 91 of the EH text.

5. BRUTELLE, Charles-Louis l'Héritier de (1746–16 April 1800).

    EH, i, 111–33; read 15 germinal an IX (5 April 1801).

    Same text in MI, iv (Hist.), 1801, 39–55.

    Separately printed as "Notice historique sur Charles-Louis l'Héritier, lue à la séance publique de l'Institut national le 15 germinal an IX" (Paris, an X, 21 pp.).

    The text of EH was also reprinted in G. Debure l'aîné, *Catalogue des livres de feu Charles-Louis l'Héritier de Brutelle* (Paris, 1802, 1805).

    The collection by Chauvin (see Section 1) also contains this study.

    Materials used by Cuvier in the writing of the EH text are to be found at BIFC 3179, 3180.

    The account in BU, xxiv, relies heavily on Cuvier's EH study.

6. GILBERT, François-Hilaire (1757–6 September 1800).

    EH, i, 137–62; read 15 vendémiaire an X (7 October 1801).
    MI, iv (Hist.), 1801, 56–73.
    Separately printed as "Notice historique sur Hilaire-François [sic] Gilbert, lue à la séance publique de l'Institut national de France, le 15 vendémiaire an X" (Paris, an X, 22 pp.).
    The EH text also appears in Gilbert's *Traîté des prairies artificielles, ou Recherches sur les espèces de plantes qu'on peut cultiver avec le plus d'avantage en prairies artificielles . . . sixième édition, augmentée de notes par M.A. Yvart . . . et précédée d'une notice historique sur Gilbert par M. le baron Cuvier* (Paris, 1826).
    The collection by Chauvin (see Section 1) also contains this study.
    Materials used by Cuvier in the writing of the EH text are to be found at BIFC 3181.
    The public reading of this *éloge* was used by Cuvier to indicate his hostility to the régime of the Consulate; see Introduction to this Bibliography, ref. 39.

7. DARCET, Jean (1725–13 February 1801).

    EH, i, 165–185; read 15 nivôse an X (5 April 1802).
    MI, iv (Hist.), 1802, 74–88.
    Separately printed as "Notice historique sur Jean Darcet, lue à la séance publique de l'Institut national le 14 nivôse an X" (Paris, an X, 19 pp.).
    The collection by Chauvin (see Section 1) also contains this study.
    Materials used by Cuvier in the writing of the EH text are to be found at BIFC 3182.
    G. Sarton, "Cuvier et les belles-lettres", *Isis*, iv (1922), 493, examines Cuvier's views on literature and science contained in this study.

8. PRIESTLEY, Joseph (1733–1804).

    EH, i, 189–234; read 5 messidor an XIII (24 June 1805). The reading date given in H. Dehérain, *Catalogue des manuscrits du Fonds Cuvier conservées à la Bibliothèque de l'Institut de France* (Paris-Hendaye, 1908–22) of 15 messidor is clearly incorrect.
    MI, vi (Hist.), 1806, 29–58.
    The EH text exists in two English translations: NEPJ, April–October 1827, 209–31; and "Historical eulogium on Joseph Priestley, read at the public sitting of the National Institute, in the Class of Mathematical and Physical Sciences, the Fifth Messidor, Year XIII, by Georges Cuvier . . . translated by the Rev. D. B. Warden M.A., and Secretary to the Minister Plenipotentiary of the United States of America, Paris" (Paris, 1807). The copy of this work in the Bibliothèque Nationale, Paris, formerly belonged to the Abbé Grégoire. Warden (1772–1845) published a translation in 1810 of Grégoire's book on the intellectual faculties of negroes. He was also a friend

of Cuvier's; Lady Morgan, *France in 1829–30* (2 vols, London, 1830), i, 456–60, contains an account of a visit to Cuvier in his company.

BU, xxxvi (1823), 83–88; a collaboration with M.-A. Pictet, who supplied most of the material on English religious life.

9. CELS, Jaques-Martin (1743–15 May 1806).

    EH, i, 237–63; read 7 July 1806.
    MI, iv (Hist.), Ser. II, 139–58.
    Separately printed (Paris, 1807, 24 pp.).
    Material for EH text is to be found at BIFC 3184.
    BU, vii (1813), 508–9.

10. ADANSON, Michel (1727–3 August 1806).

    EH, i, 267–308; read 5 January 1807 (Dehérain gives 4 January).
    MI, vi (Hist.), Ser. II (1806), 159–88.
    Separately printed as "Eloge historique de M. Adanson, lu à la Classe des sciences mathématiques et physiques de l'Institut national dans la séance publique du 5 janvier 1807" (Paris 1807, 32 pp.).
    Material used by Cuvier in the writing of this *éloge* is to be found at BIFC 3185/6; p. 272 of EH refers to Cuvier's reliance on Adanson's own papers, including his autobiography. EH text of 1819 was revised in punctuation and spelling from that of 1807.
    This study, which contains one of Cuvier's most important discussions of classification in the natural sciences, was also reprinted in all editions of Cuvier (ed. Flourens), and in the *Nouvelle bibliothèque populaire* (see Section 1).
    Translated into English by Robert Jameson, NEPJ (April–October 1827), 1–21.

11. BROUSSONET, Pierre-Marie-Auguste (1761–21 July 1807).

    EH, i, 311–42; read January 1808.
    MI, vii (Hist.), Ser. II (1807), 93–115.
    Material used by Cuvier in preparing this *éloge* is to be found at BIFC 3186.

12. LASSUS, Pierre (1741–16 March 1807).

    EH, i, 345–60; read 2 January 1809.
    MI, viii (1808), Hist., 85–96.
    Material used by Cuvier in the preparation of this study is to be found at BIFC 3187.

13. VENTENAT, Etienne-Pierre (1757–13 August 1808).

    EH, i, 363–79; read 2 January 1809.
    MI, viii (1808), Hist., 97–108.

Material used by Cuvier in the preparation of this *éloge* is to be found at BIFC 3188.

14. BONNET, Charles (1720–20 May, 1793).

EH, i, 383–430; read 3 January 1810. Cuvier was absent in Italy.
This *éloge* of Bonnet was included in that of Saussure; the pages immediately relating to Bonnet are 386–409.
Materials for the construction of both *éloges* are to be found at BIFC 3189.
It was translated by Robert Jameson, NEPJ (1827), pt ii, 213–35.
BU, v (1812), 130–2.
In both EH and BU, Cuvier defends Bonnet from charges of materialism, and does not quarrel with his support of the theory of the pre-existence of germs.

15. SAUSSURE, Horace Bénédict (1740–22 January 1799).

EH, i, 383–430, and especially 410–30 (see comments on no. 14); read 3 January 1810.
Sources for EH are to be found at BIFC 3190.
BU, xl (1825), 476–9.

16. FOURCROY, Antoine-François (1753–16 December 1809).

EH, ii, 3–53; read 7 January 1811.
MI (1810), pp. xcvi–cxxviii.
AMNHN, xvii (1811), 99–131.
Magazin encyclopédique, 1812, II, 1–32.
Separately printed as "Eloge historique d'Antoine Fourcroy, lu à la séance publique de l'Institut national le 7 janvier 1811" (no place or date, 34 pp.).
The EH text was also reprinted in Flourens and Chauvin (see Section 1).
BU, xv (1816), 367–71.

17. AGRICOLA, George (1494–1555).

BU, i (1811), 311.

18. ALDROVANDI, Ulysse (1527–1605).

BU, i (1811), 474.
Cuvier had obviously seen the remnants of Aldrovandi's collection in Bologna.

19. ARISTOTLE.

BU, ii (1811), 456–64.

20. CAVENDISH, Henry (1731–24 February 1810).

EH, ii, 77–105; read January 1812.
MI, 1811, pp. cxv–cxxv.
NEPJ, April–September 1828, 209–22.
BIFC 3193.

21. Desessarts, Jean Charles (1729–16 April 1811).

    EH, ii, 57–73; read 6 January 1812.
    MI, 1811, cxv–cxxv.
    BIFC 3192.

22. Boyle, Robert (1627–30 December 1691).

    BU, v (1812), 427–31.

23. Buffon, Georges Louis Leclerc, Comte de (1707–16 April 1788).

    BU, vi (1812), 234–42.
    Reprinted in *Histoire naturelle de Buffon, mise dans un nouvelle ordre, précédée d'une notice sur la vie et les ouvrages de cet auteur par M. le baron Cuvier* (36 vols, Paris, 1825–26).
    Oeuvres complètes de Buffon (Paris, 1830).
    Oeuvres complètes de Buffon, précédées d'une notice sur la vie et les ouvrages de Buffon par le baron Cuvier (Paris, 1830–32).
    C. Hemardinquer (ed.), Buffon, *Discours sur le style* (Paris, 1877, 1886, 1897); see Appendix in all these editions.

24. Pallas, Peter Simon (1741–8 September 1811).

    EH, ii, 109–56; read 5 January 1813.
    NEPJ, April 1827–October 1828, 211–37.
    BIFC 3194; 3242/47.

    The article in BU, xxxii (1823), 437–88, is based on the EH text.

25. Comparetti, Andreo (1746–22 December 1801).

    BU, ix (1813), 371–3.

26. Daudin, François-Marie (*c.* 1774–1804).

    BU, x (1813), 572–4.

27. Dolomieu, Déodat-Gui-Sylvain-Tancrède de Gratet de (1750–26 November 1801).

    BU, xi (1814), 493–9.
    BIFC 3148.

28. Edwards, George (1693–1773).

    BU, xii (1814), 542–4.

29. Parmentier, Antoine-Augustin (1737–17 December 1813).

    EH, ii, 159–231 (included with a study of Rumford); read 9 January 1815.
    Reprinted in the collected editions by Chauvin and Flourens (see Section 1).

30. RUMFORD, Benjamin Thompson, Count (1753–21 August 1814).

EH, ii, 159–231 (included with a study of Parmentier); read 9 January
  1815.
Translated by Jameson, NEPJ (January–April 1830), 209–29.
BIFC 3195.
EH does not mention Rumford's marriage to Mme Lavoisier.

31. OLIVIER, Guillaume-Antoine (1756–1 October, 1814).

EH, ii, 235–65; read 8 January 1816.
Separately printed as "Institut de France, séance publique du 8 janvier 1816,
  Discours sur Guillaume-Antoine Olivier".
Reprinted in E. Olivier, *G. A. Olivier, sa vie, ses travaux, ses voyages, docu-
  ments inédits* (Moulins, 1880).

32. FLANDRIN, Pierre (1752–96).

BU, xv (1816), 22–23.

33. FONTANA, Felix (1730–9 March 1805).

BU, xv (1816), 196–9.
Much of the material for this article can be traced to Cuvier's correspondence
  with Giovanni Fabroni, Fontana's subordinate in Florence.

34. GESNER, Conrad (1516–65).

BU, xvii (1816), 242–7.

35. TENON, Jacques-René (1724–18 January 1816).

EH, ii, 269–304; read 17 March 1817.
BIFC 3196, including Tenon's autobiography.
Reprinted in Chauvin (see Section 1).

36. GUYTON DE MORVEAU, Louis-Bernard (1737–2 January 1816).

BU, xix (1817), 262–5.
BIFC 3144.

37. HALLER, Albrecht von (1708–12 December 1777).

BU, xix (1817), 330–7.
BIFC 3152.

38. HERMANN, Johann (1738–4 October 1800).

BU, xx (1817), 257–60.
BIFC 3147; 3239/8; 3252/77.
Much of Cuvier's information for this article came from Hermann's brother,
  Johann Frederick, whom Cuvier had first met in Stuttgart.

39. DESMAREST, Nicolas (1725–20 September 1815).

EH, ii, 339–74; read 16 March 1818.
BIFC 3141/1; 3199 contains a biography of Desmarest's son Anselme-Gäetan.
Cuvier also delivered a graveside oration, separately printed as "Institut royal de France. Funerailles de M. Desmarest, le 29 septembre 1815. M. le Chevalier Cuvier a prononcé le discours suivant" (Paris, 1815, 3 pp.).
K. L. Taylor, "Nicolas Desmarest and geology in the eighteenth century", in C. J. Schneer (ed.) *Toward a history of geology* (Cambridge, Mass., 1969), 339n., indicates the extent to which presentation of Desmarest has been dominated by Cuvier's account.

40. WERNER, Abraham Gottlob (1750–30 June 1817).

EH, ii, 307–36; read 16 March 1818.
BIFC 3197; 3198. The latter has been commented on in W. Coleman, "Abraham Gottlob Werner vu par Alexandre von Humboldt avec des notes de Georges Cuvier", *Sudhoffs Archiv*, lvii (1963), 465–78.
EH was translated by Robert Jameson, NEPJ, v (1821), no. vii, 1–16; and a different translation was also reprinted in James Duncan, *British butterflies* (London, 1835, 1876).
BU, 1 (1827), 376–9.

41. JONSTON, John (1603–75).

BU, xxi (1818), 631–2.
BIFC 3151 shows that this article was in fact largely written by Cuvier's brother Frederick.

42. LAMANON, Robert de Paul (1752–10 December 1787).

BU, xxiii (1819), 255–6.

43. LAVOISIER, Antoine-Laurent (1743–8 May 1794).

BU, xxiii (1819), 461–7.
BIFC 3145.
Cuvier devotes over half the length of this article to a study of Lavoisier's administrative career; one suspects here the influence of information from Mme Cuvier, whose first husband was not only executed on the same day as Lavoisier as another *Fermier-Général*, but also worked with him in the Société des Amis des Noirs. See L. Cahen, "La Société des Amis des Noirs et Condorcet", *La Révolution française*, lx (1906), 481–511.

44. LINNAEUS, Karl von (1707–10 January 1778).

BU, xxiv (1819), 525–35.

45. PALISOT DE BEAUVOIS, Ambroise-Marie-François-Joseph (1755–21 January 1820).

EH, iii, 3–45; read 27 March 1820.

MAS, iv (1819–20), Hist., 318–44.
BIFC 3200, 3201.
Translated by Robert Jameson, NEPJ (October 1828–March 1829), 1–21.

46. LYONNET, Pierre (1707–10 January 1789).

BU xxv (1820), 532–4.
BIFC 3150.

47. BANKS, Sir Joseph (1743–19 March 1820).

EH, iii, 49–92; read 2 April 1821.
MAS, v (1821–22), Hist., 204–30.
MMNHN, xiii (1825), 297–326.
Separately printed as "Eloge historique de Sir Joseph Banks, lu le 2 avril 1821".
Translated by Robert Jameson, NEPJ (1827), 1–22.
In a different translation, the EH text is reproduced as the introduction to George Suttor (ed.), *Memoirs historical and scientific of the Right Hon. Sir Joseph Banks, Bart* (London, 1855).
BIFC 3202.

48. LA METHERIE, Jean Claude de (1743–1 July 1817).

BU, xxviii (1821), 461–3.

49. MULLER, Othon-Frederick (1730–26 December 1784).

BU, xxx (1821), 394–6.

50. DUHAMEL, Jean-Pierre-François-Guillot (1730–19 February 1816).

EH, iii, 95–120; read 8 April 1822.
MAS, vi (1823), Hist., 160–76.
BIFC 3203.
NEPJ (October 1830–April 1831), 1–14.

51. VAN SPAENDONCK, Gérard (1746–11 May 1822).

EH, iii, 435–40; read 13 May 1822.
Separately printed as "Institut royale de France. Académie royale des sciences. Funerailles de M. van Spaendonck".
A graveside oration, rather than a fully-developed *éloge*.

52. DELAMBRE, Jean-Baptiste-Joseph (1749–19 August 1822).

EH, iii, 441–7; read 21 August 1822 (a graveside speech rather than a fully developed *éloge*).
Separately printed as "Institut royale de France: Académie royale des sciences. Funerailles de M. le Chevalier Delambre" (Paris, 1822, 6 pp.).

53. HAUY, René-Just, Abbé (1743–3 June 1822).

> EH, iii, 123–75; read 2 June 1823.
> MAS, viii (1829), Hist., 145–77.
> MMNHN, x (1823), 1–35.
> NEPJ (October 1828–March 1829), 205–31.
> Also reprinted in Chauvin and Flourens (see Section 1).
> Cuvier also pronounced a short graveside oration: "Institut royale de France. Académie royale des sciences. Funerailles de M. l'Abbé Haüy . . . M. le baron Cuvier a prononcé, au nom des deux établissements, le discours suivant" (Paris, 1822, 4 pp.); read 3 June 1822.

54. FABRONI, Giovanni-Valentino-Matteo (1752–17 December 1822).

> EH, iii, 405–34; no reading date.
> BIFC 3213.

55. PICOT DE LA PEIROUSE, Philippe (1744–18 October 1818).

> BU, xxxiii (1823), 262–4.
> BIFC 3146.

56. PENNANT, Thomas (1726–16 December 1798).

> BU, xxxiii (1823), 315–18.

57. PLINY, (the elder).

> BU, xxxv (1823), 67–76.

58. BERTHOLLET, Claude-Louis (1748–6 November 1822).

> EH, iii, 179–227; read 7 June 1824.
> MAS, viii (1829), Hist., 179–210.
> Separately printed in "Académie des sciences, recueil des discours lus dans la séance . . . du 7 juin 1824" (Paris, 1824).
> BIFC 3206.

59. RAY, John (1628–17 January 1705).

> BU, xxvii (1824), 155–63 (in collaboration with Aubert Dupetit-Thouars). Translated by G. Busk, as preface to E. Lankester, *Memorials of John Ray* (London, 1846). This volume, printed for the Ray Society, also contains memoirs by Dr Derham and Sir J. E. Smith.

60. REAUMUR, Réné-Antoine-Ferchault de (1683–18 October 1757).

> BU, xxxvii (1824), 198–203.

61. ROESEL, Auguste-Jean (1705–27 March 1759).

> BU, xxxviii (1824), 398–9.

62. RONDELET, Guillaume (1507–66).

    BU, xxxviii (1824), 546–9.

63. RICHARD, Claude-Louis (1754–7 June 1821).

    EH, iii, 231–57; read 20 June 1825 (reading date given in MAS).
    MAS, vii (1827), Hist., 195–212.
    MMNHN, xii (1825), 359–66.
    BIFC 3207.
    NEPJ (July to October 1830), 201–14.

64. THOUIN, André (1747–23 September 1824).

    EH, iii, 261–78; read 20 June 1825.
    MAS, vii (1827), Hist., 213–24.
    MMNHN, xiii (1825), 205–16.
    BIFC 3208.
    Graveside speeches, separately printed as "Académie royale des sciences.
        Funerailles de M. Thouin, discours par M. le Baron Cuvier et P. L. A.
        Cordier" (Paris, 1824, 8 pp.). This speech was also printed in MMNHN,
        xii (1825), 1–5.

65. SCHEUCHZER, Jean-Jacques (1677–1733).

    BU, xli (1825), 116–19.
    Interesting references to Cuvier's own examination of the *Homo diluvii testis*.

66. SCHNEIDER, Jean-Gottlob (1750–13 January 1822).

    BU, xli (1825), 199–206.
    Schneider's edition of Aristotle's *Natural history* (4 vols, Leipzig, 1811), was
        dedicated to Cuvier.

67. SEBA, Albert (1665–1736).

    BU, xli (1825), 399–401.

68. LACEPEDE, Bernard-Germain-Etienne de la Ville, Comte de (1756–6 October
    1825).

    EH, iii, 281–335; read 5 June 1826.
    MAS, viii (1829), Hist., 212–48.
    MMNHN, xiii (1825), 369–404.
    BIFC 3209; including Lacépède's autobiography.
    Other printings:
    *Histoire naturelle de l'homme, par M. le comte de Lacépède, précédée de son
        éloge historique par M. le baron Cuvier* (Paris, 1827, 1839).
    *Oeuvres du comte de Lacépède, comprenant l'histoire des quadrupèdes ovi-
        pares, des serpents, des poissons, et des cétécés* (13 vols, Paris, 1830).

The *éloge* by Cuvier is in vol. xiii. This work was reprinted in
1839, 1855, 1856, 1860.
*Académie des sciences, recueil des discours lus dans la séance du
5 juin 1826* (Paris, 1826).

69. CORVISART, Jean-Nicolas des Marets, Baron (1755–18 September
1821).

EH, iii, 361–78 (with Hallé and Pinel); read 11 June 1827.
MAS, ix (1830), Hist., 197–260.
NEPJ (January–April 1830), 1–18.
BIFC, 3211.
Also printed as: *Séance publique de la Faculté de médecine de
Paris du 22 novembre 1821. Discours de Duputren et de
Cuvier sur Corvisart* (Paris, 1821, 56 pp.).

70. HALLE, Jean Noël (1754–1822).

EH, iii 345–60 (with Pinel and Corvisart); read 11 June 1827.
MAS, ix (1830), Hist., 197–260.
BIFC 3210.
NEPJ (January–April, 1830), 1–18.

71. PINEL, Philippe (1745–25 October 1826).

EH, iii, 379–402 (with Hallé and Corvisart).

72. VICQ D'AZYR, Félix (1748–20 June 1794).

BU, xlviii (1827), 374–8 (not vol. xliii, as stated by Dehérain).
BIFC 3149.

73. WALLERIUS, Jean Gottschalk (1709–85).

BU, 1 (1827), 127–9.

74. RAMOND, Louis-François-Elizabeth de Carbonnières, Baron
(1755–14 May 1827).

Separately printed as "Institut royal de France, Académie royale des sciences. Séance publique du lundi 16 juin 1828. Eloge historique
de Louis-François-Elizabeth baron Ramond" (Paris, 1828, 27 pp.).
MAS, ix (1830), Hist., 169–95.
BIFC 3154; 3250/43 contains a letter from the baronne Ramond
demanding modifications in Cuvier's text, of 30 July 1828;
however, the MAS text presents no substantial changes.
See also [H. Dehérain], "Une autobiographie du baron Ramond",
Journal des savants (March 1905), 121–30; reprints Cuvier's
major source for his study of Ramond, the autobiography of
February 1827.

75. Bosc, Louis-Auguste-Guillaume d'Antic (1759–10 July 1828).

   MMNHN, xviii (1829), 69–92.
   MAS, x (1831), Hist., 191–218.
   Separately printed as "Institut royal de France. Eloge historique de M. Bosc,
      lu à la séance publique de l'Académie royale des sciences le 15 juin
      1829" (Paris, 1829, 28 pp.).
   The Bibliothèque Nationale, Paris, possesses another edition of this pam-
      phlet, of 24 pp., no place or date.
   NEPJ (April–October 1829), 274–80.
   Cuvier also delivered an oration at Bosc's funeral, 12 July 1828, which con-
      tains a complimentary reference to Lamarck.
   BIFC 3157, including Bosc's autobiography.
   The MMNHN gives the reading date of the funeral oration as 15 June 1828,
      but the MAS and separate printing reading date is obviously the correct
      one.

76. Daru, Pierre-Antoine-Noël, Comte (1767–5 September 1829).

   MAS, xii (1833), Hist., 1–37; read 11 September 1829 [graveside speech].
   BIFC 3141/2.
   Separately printed as: "Discours prononcés aux funerailles de M. le baron
      Daru: Discours de M. Cuvier: Institut royal de France: 11 septembre
      1829" (Paris, 1829, 6 pp.).

77. Fourier, Jean-Baptiste-Joseph, Baron (1768–16 May 1830).

   A graveside speech: "Discours prononcés aux funerailles de M. le baron Fou-
      rier: Discours de M. le baron Cuvier, 18 mai 1830" (Paris, 1830, 3 pp.).
78. Davy, Sir Humphry (1778–29 May 1829).

   MAS, xii (1833), 1–37; read 26 July 1830.
   Separately printed, December 1832.
   BIFC 3158 contains letters between Lady Davy and Cuvier on the subject of
      the *éloge*.
   J. Z. Fullmer, "Davy's biographers: notes on scientific biography", *Science*,
      clv (1967), 285–91, mentions Cuvier's study only briefly, and it is com-
      pletely ignored by both major contemporary biographers, Paris and John
      Davy, mentioned by Fullmer.

79. Vauquelin, Nicolas-Louis (1763–14 November 1829).

   MAS, xii (1833), pp. xxxix–lvi; read 26 July 1831.
   BIFC 3153.
   Originally designed to be read after that of Davy.
   Reprinted in Chauvin (see Section 1).

80. Lamarck, Jean-Baptiste-Pierre-Antoine-de Monet, Chevalier de (1744–18
   December 1829).

MAS, xiii (1835), 1–31; read 26 November 1832; Cuvier had died in May. BIFC 3156.

Originally written for the *séance* of 27 June 1831, to follow that of Volta by Arago (see Section 3 for the ascription of this *éloge*).

BIFC 3252/47 contains a letter from Lamarck's son to Cuvier of 8 February 1830, heavily insisting on the 'solitude' element in the mythology of Lamarck: "Tous les jours de mon père se ressemblaient: ne sortant jamais de son cabinet, il s'en était fait un monde dont ses livres et ses collections étaient les seules inhabitants."

E. T. Hamy, *Les débuts de Lamarck* . . . (Paris, 1908), 3–17, prints a further letter from Lamarck's son to Cuvier of 20 February 1830.

The best recent study of this notorious *éloge* is Robert Courrier, "Georges Cuvier (1769–1832), certains aspects de sa carrière", *Institut de France: notes et discours*, v (1963–72), 641–61.

C. Doubtful *éloges*

Chiefly works mentioned in the literature on Cuvier as being by him, obviously wrongly ascribed or of which no further evidence can be found.

1. EH, iii, 343, mentions projected future *éloges* of Sabatier, Percy and Deschamps.
2. L. Marchant (ed.), *Georges Cuvier, lettres à C. H. Pfaff sur l'histoire naturelle, la politique et la littérature, 1788–1792* (Paris, 1858), 37, quotes Pfaff's ascription of an *éloge* of Alessandro Volta, in fact written by Arago.
3. Cuvier began work on studies of Young and Wollaston, which were uncompleted at the time of his death. Their *éloges* were finally written and delivered by his successor as Perpetual Secretary, François Arago. Cuvier's materials for the projected studies are to be found at BIFC 3143; 3252/39, 40, contains letters from Hudson Gurney to Cuvier of 17 and 23 August 1830 enclosing documents on Thomas Young for Cuvier's forthcoming *éloge*.
4. S. Moravia, *Il pensiero degli idéologues: scienza e filosofia in Francia 1780–1815* (Florence, 1974), 581, notes 115, 116, ascribes to Cuvier an "Eloge de François Péron lu à la classe des sciences physiques et mathématiques, le 9 juillet 1806", and continues: "Questo testo è stato ripubblicato anche nel *Voyage* dello stesso Péron, vol. I, pp. 1–18 della II ed. dell'opera (Paris, 1824)." Apart from the fact that Péron did not die until 1810, I have been unable to locate any version of the text of this *éloge*.
5. Cuvier was commissioned to write an article for BU, vol. 1, on the English naturalist Woodward; however, his contribution was rejected as unsatisfactory by its editor, Michaud, and does not survive in BIFC. See BIFC 3249/31, Michaud to Cuvier of 22 December 1827.

6. The same fate seems to have overtaken Cuvier's projected article on Spallanzani for BU. Cuvier's name appears on the signature page of BU, vol. xlii (1825), but in fact there is no article by him in this volume, and it seems almost certain that this article was either rejected, or was incomplete by the printing date of this volume of BU. See BIFC 3255/16, Michaud to Cuvier, no date.

D. Alphabetical index of biographical subjects

The numbers given to each *éloge* refer to the chronological list above.

3

POLITICS AND VOCATION

French science, 1793–1830

From: *British Journal for the History of Science*, 13 (1980), 27–43. © Cambridge University Press. Reprinted with permission.

French science of the period between 1793 and 1830 is now a major focus of study. The large body of work produced since the nineteenth century, particularly in the field of institutional history,[1] has provided the background for important attempts in the last ten or fifteen years to apply tools of sociological analysis to this field of enquiry. Particularly important have been theories of professionalization and institutionalization.[2] It is the purpose of this paper to examine the consequences of the use of such models in relation to this specific historical context. In particular, I shall suggest that such questions as the importance of institutions in the conduct of science, and the extent to which science became a profession or remained a vocation, may be better understood once the world of French science has been situated in a wider political and intellectual context. An article, however, can do no more than suggest new perspectives, and must leave to more extended treatments the work of amplification and correction. Briefly, however, this paper will argue for a view of science at this period as locked in a conflict between the ambiguous demands of the political world on the one hand, and on the other pressures on individuals and groups within the vocation of science to conform to an ideology which viewed science as completely *non*-political.

The wide range of models of professionalization has been usefully reviewed recently,[3] and it would serve no purpose to offer here full summaries of all such theories. Though there is no single or simple definition of professionalization, all models contain some at least of the following stipulations; that the occupation concerned should be remunerated and followed full-time; that it should be entered only after specialized training given by other professionals, and certified by some form of impersonal examination; and that it should be followed in accordance with a high standard of competence, again impersonally maintained and defined by other professionals. Today, the enforcement of professional standards is internal to the field in question, and is concentrated in institutions concerned with the delineation of the field of competence of that particular profession.

DOI: 10.4324/9781003038085-4

However, many of these criteria are difficult to apply to science in the period under discussion. They assume, first, that fields of competence can be readily distinguished from one another, and that it is seen as normal to make such distinctions. However, recent work on the history of scholarship has made it clear that such ways of defining activity were not fully worked out by the beginning of the nineteenth century, and that the process took even longer to be accepted in the sciences than in the humanities.[4] Nor did institutions set standards in the sciences; for example, few degrees in general science were awarded in this period. The individuals who dominated the French scientific world, men such as Cuvier, Laplace, Berthollet, Biot, and Geoffroy St Hilaire, possessed no formal qualifications in science.[5] Not until the very end of this period did institutions begin to fulfil the functions demanded by the models of professionalization. This is not to say that they were not important; rather that the nature of their importance needs to be redefined.

A further difficulty in the application of models of professionalization to the science of this period arises from the fact that the traditional professions such as medicine and the law assume an applied component, an ideal of service. In the natural sciences, however, it is difficult to identify the client served or the service provided. This is especially so when we look at the period before the rise of state-funded 'big science'. Science, in the sense of a specific form of culture concerned with enquiry into the natural world, cannot readily be defined in terms of the vocabulary of professionalization.

Functionalist sociology has laid a special emphasis on the role of institutions, which are seen as crucial anvils 'on which the often conflicting values of science and society are shaped into a viable force'.[6] Institutions are instruments of the 'needs' of science, a science thought of very largely as an autonomous entity, whose intrinsic value-freedom must be protected from such bearers of subjective external pressures as systems of political power. Scientific institutions guarantee the autonomy of science, and hence its authenticity as science. Even this crude and abbreviated summary of functionalist analysis indicates the extent to which the institutional life of science in historical periods has been treated as the defining form of that life; it should also have made clear how the ideology of science as a vocation, whose autonomy is guaranteed by abstention from the political realm, has been carried over into the account of how such institutions function in the real world.

A more sharply focused account of science in this period must begin with a comparison of the periods before and after 1793.[7] There were in fact many institutions connected with the teaching or advancement of science before 1793, even though the number of people involved in them remained tiny compared to the total population of France.[8] 'La physique' was taught in some form in most of the secondary schools run by the religious orders, and especially in those of the Oratorians. The twenty-two French universities existing in 1791 (with a total enrolment of about five thousand students) possessed between them ten medical faculties engaged in active teaching. Special schools for naval doctors existed at

Brest, Toulon, and Marseille. Anatomical instruction was offered at the Collège royal and at the Jardin du roi in Paris, and in many private courses offered by members of the medical faculty of the capital. Mathematics was offered at the universities, at the training schools for artillery officers (about a hundred teaching posts for the whole of the eighteenth century) and at the civil engineering school of the Ponts-et-chausées (whose maximum enrolment was just under a hundred pupils). Botany and chemistry could be learnt at most of the universities offering medical courses, as well as at the Jardin du roi. With the exception of the Collège de pharmacie in Paris, pharmacy remained a speciality rarely taught outside trade apprenticeship. The royal observatory in Paris continued to dominate astronomy, while the establishment of the Ecole des mines in 1783 brought new opportunities for the study of geology.[9]

It is not clear that the provision of institutions either of teaching or research in science improved dramatically after 1793. The property of the teaching orders was confiscated by the revolutionary government after 1792, and their schools thus forced to close. In spite of the strong emphasis on the teaching of science of their replacements, the state-run *écoles centrales*, one per *département*, these could not hope to equal the former level of provision in secondary education. They were in any case abolished themselves in 1802. Their successors, the *lycées*, again run by the state, were far more socially exclusive in their intake, and on Napoleon's own orders reduced their scientific instruction to the mathematics needed to compete for entry into the Ecole polytechnique.[10] The university structure abolished in 1791 was not re-established until the foundation of the Imperial University in 1808. Even then, provision for the teaching of science could hardly be said to have been overwhelmingly generous in relation to that for other subjects.[11] Nor do the institutions of more advanced scientific learning located in the capital present a picture of massive expansion. The number of posts they offered was small,[12] and the real availability of posts was further reduced by two other factors: by the dominance of certain families within given institutions,[13] and by the practice of pluralism.[14]

If the scientific institutions of Napoleonic France offered no dramatic improvement on those of the *ancien régime*, neither do they satisfy modern expectations about the role of the institutions in 'professionalized' science. Institutional employment in science at this period could certainly not necessarily be taken as a validation of competence in science; nor was the availability of posts such that it was possible to use institutions to provide a regular *cursus* of employment for those potentially qualified for it.[15] Institutions also faced continual financial problems, the priority of the army in claims for the resources of the Napoleonic state was always clear.[16] Institutionalization was not synonymous with autonomy, security, or continuity. Nor was the Napoleonic government strongly committed to the advancement of science as such. As general policy became increasingly conservative after the Concordat of 1802, educational policy again endorsed a classical literary culture, with immediate effects on the *lycée* curriculum.[17] The science teaching available at the Ecole polytechnique was within a few years of its foundation deliberately restricted by Napoleon to a technical programme of mathematics and engineering. Other new institutions, such as the schools of pharmacy founded in 1803 at Paris, Strasbourg, and Montpellier, had a strongly technical bias.

Analysis of the institutional structure of science can indeed be helpful to the historian. It gives an indication of the physical and financial provision for certain kinds of enquiry. It defines some of the conditions of employment in science, and the possibility of income from it. It reveals some of the rewards and penalties which scientific patrons with secure institutional tenure can dispense to their protégés; it thus also says something about the ability of such patrons to maintain research programmes. Analysis of gross institutional tendencies may also reveal, as the French case certainly does, a bias in the kind of support given to science by the state, which in turn may generate competing factions within the scientific community.

However, institutional analysis cannot answer all the questions we might want to ask. Institutions did not define the scientific community and its entrants by means of examination. The lack of a clear distinction in this period between amateur and professional – and between scientist and humanist – also makes it difficult to use institutional affiliation as a measure of contribution to science. There are also many questions about institutions themselves that we still find it difficult to answer. We know little, for example, about the experience of students of science and medicine during the Empire and the Restoration. What was the effect on science in general of the undoubted increase in the numbers of those taught medicine or some form of applied science? How did this development affect the public image of science, or increase tendencies towards scientific specialization and professionalization?[18] How far was the institutional structure of science during the Restoration able to absorb potential aspirants to it?[19]

The list of questions could be extended. For their answer, these questions require a knowledge both of the inner life and of the public role of science which we do not as yet possess in coherent form. One of our main difficulties arises out of the uncertainties of contemporaries themselves. Intellectual élites as such had only recently achieved a place in public life when the collapse of government by monarchy and aristocracy had left the intellectual class as one of the favourite reservoirs of talent in the service of the hard-pressed revolutionary state. But the nineteenth century was no more successful than the *ancien régime* in the resolution of what Daniel Roche has described as one of the most pressing problems faced by any society, that of the definition of its real élite.[20] Scientists were particularly difficult for contemporaries to place in social terms, because of the very insistence of the ideology that true science was an *a*political, *a*social activity: the true scientist withdrew from human society and its corrupting influences in order to gain the closest possible contact with nature. Scientists who wielded power in the real political world, or who owed their ability to exercise their vocation to government subsidy, were thus in a social position difficult to define in terms of vocation alone.[21]

To a lesser extent, of course, the traditional professions also encountered these conflicts between internal and external reference points in their placing in

society. The legal profession, for example, had experienced this tension of definition acutely before the Revolution.[22] Successful lawyers defined their capacity
to rise in society as the capacity to buy into the nobility and adopt its life-style;
yet professional eminence, the motor of their social ascent, was simultaneously
defined by the standards of professional excellence prevailing amongst their fellow practitioners. There seemed to be few ways of reconciling these two competing definitions.

Even so, the traditional professions could rely, to a far greater extent than could
the sciences, on public acceptance in a more general sense. Doctors, for example,
could rely on the widespread endorsement of the 'ideology' of their profession of
service to the community, and on public and family approval of individuals who
sought to become doctors. The 'ideology' of natural science, on the other hand,
approached the actual social world with far less certainty. The objectivity essential for correct observation of the natural world could be gained, it was felt, only at
the price of withdrawal from the social and political arena. Further, the choice of
scientific career was usually accompanied by parental hostility, physical hardship,
and financial insecurity.[23] Only government seemed to be able to give a positive
endorsement to the public life of non-applicable science, by conferring prestige
through the employment of scientists in state institutions. Strong personal relationships or working partnerships with Napoleon on the part of scientists such as
Monge, Berthollet, and Chaptal, and the employment of such men in high political positions, further increased the public status of the scientific intellectual – just
as such employment and such relationships of the literary figures Antoine Arnault
and Louis de Fontanes helped to validate the entry of the intellectual class as a
whole into public life. During periods of rapid change, intellectuals of all kinds
become useful to governments because of their ability to manipulate former rules
and formulae in new directions.

But the new French state was still marked by many uncertainties. Composed
of citizens instead of subjects, and claiming to represent the highest good of its
citizens rather than divine support for the rule of one man and his family, the
new state could offer public validation to individuals and groups to an extent
far beyond that possible with the more limited claims of the *ancien régime*. But
the groups to be validated by the new state were almost too various; there were
almost too many definitions of who was qualified to take on a public role. Napoleon was engaged in a policy of reconciliation with the aristocracy of the old
regime. He was also trying to increase the acceptability of employment by the
state. New technical professions such as engineering and pharmacy were also in
the process of formation. Other occupational groups, such as soldiers, and savants
of all kinds,[24] were especially favoured in the distribution of office. By personal
intervention, such as by his manifestations of public interest in the life of the Institut de France, Napoleon underwrote the credentials of this new group to occupy
positions which usually had little obvious connexion with their vocational affiliation. Many former soldiers intervened in local administration, for example, while
Chaptal's tenancy of the Ministry of the Interior seemed to have little to do with

the prosecution of his chemical research.[25] This is something of what contemporaries meant when they spoke of the accession of the sciences to 'une existence politique' after the Revolution.[26]

The Napoleonic political arena, in fact, had to be loosely enough defined to contain all the competing definitions of élite status that Napoleon's many different concurrent policies towards the formation of his ruling class created. The political arena was much less strongly defined than is usual in modern Western societies. Politics, bureaucracy, and intellectual life flowed into each other; individuals, methods, attitudes, and metaphors were transferable over the whole area.[27] What did this mean about the kinds of power at the disposal of the dominant figures of French science? Power, first of all, tended to be highly personalized. Dominance came not only from institutional affiliation, but also from the construction of systems of personal patronage which mobilized clients and services over a wider range than that provided by a single institution. Both Laplace and Berthollet, for example, achieved success, reputation, and pupils by establishing a patronage network based on their own personal and peculiar accumulation of different kinds of power – power through institutional affiliation, through their relationship with Napoleon, through financial independence based on state employment, and power to influence the organs which disseminated scientific knowledge, such as journals and the forum of the Institut.[28]

The powerful patron was crucial to the shaping of career patterns. In the case of Biot, for example, the acquisition of Laplace and Lagrange as patrons long preceded, and was prerequisite for, the obtaining of a job in Paris.[29] In the case of Geoffroy St Hilaire, a close personal relationship with René-Just Haüy, his former teacher, and then with Daubenton, were prerequisites for his appointment at the Museum at the age of twenty.[30] The power of each patron was produced by a complex interaction between his personal prestige, his success in conforming to ideals of the good scientist, and the specific opportunities represented by each of the institutions with which he was connected.[31] It was enhanced by the particular web of social connexions with which he could surround himself. Above all, it was the patron who guaranteed the worthiness of his client to enter the scientific community, but we still lack both a generalized theory of patronage for this period, and a systematic listing of the *gestes* and conventions which defined entry and acceptance into the scientific community.

Just as theories of institutionalization and professionalization can be of only limited use to our understanding of French science in this period, so sociological analysis of modern political systems of patronage proves also to be of limited applicability.[32] Such analyses tend to be concerned with patrons who act as brokers between the state and a 'backward' sector of society. They rarely envisage the case of a client achieving equality of status with his original patron. Yet in the case of Napoleonic and Restoration science, this is precisely what the patron offered as his service to the client. In these recent analyses too, the identification of the patron is often taken as being self-evident. But in a world where areas of power were weakly defined, and where personal relationships were the major pathways of influence, this was not necessarily the case. The historian needs to be able to list criteria in virtue of which contemporaries were able to recognize individuals as likely patrons. Institutional affiliation provides one very important set of indicators, and can be readily established from official lists of office-holders such as the *Almanach impérial* and its successor, the *Almanach royal*. It is much more difficult, however, to document the social and family contacts upon whose exploitation successful patronage also depended. There were for example individuals interested in science, but perhaps without affiliation to a scientific institution, whose social contacts would nevertheless be extensive enough to make them valuable intermediaries between an aspiring client and a choice of potential patrons who did possess office.[33] These social aspects of patronage have to be reconstructed from a variety of literary sources. First there is the contemporary memoir;[34] membership lists of private cultural and philanthropic societies, and of contributors to journals, are also accessible sources.[35] From systematic use of biography and autobiography,[36] it is possible to reconstruct the patterns of approach used by clients to potential patrons. Accounts of behaviour in the *salons* of the great patrons are abundant, but have never been fully analyzed as a repertoire of forms of social action.[37] In particular, we are concerned with the ways in which the patron's approval might be signified; the ways in which ordinary social actions, such as conversation and gossip, might be used to fix reputation and acceptability;[38] and the ways in which the patron could communicate to the aspirant his arrival amongst a community of his peers.[39] These explorations would tell us much about the way the evaluation of vocation was socialized, since the individual patron was the meeting point between science as a vocation and science as a political system. It would also tell us much about the forms of sociability amongst the élite as a whole, and thus allow a clearer idea to be formed of the distinction between public and private realms in this period.

We also need more precise knowledge of the way in which the state validated individuals and groups. Unlike the case of the liberal state which was to succeed it,[40] patronage was the dynamic of Napoleonic government. State employment validated a position as a patron. The Institut presents a good example of the way the government could confer prestige and endorsement through the use of an institution. Rather like Louis XIV's Versailles, the Institut provided both a grouping point for a large segment of the politico-cultural élite, and an arena where the ability of the state to inflict disgrace or award prestige could be applied in exemplary ways. Pressure from government on such an institution was correspondingly severe. Conflicts amongst its members over election to vacant places meant that the Institut had in any case become as much a showpiece of division amongst the intellectual community as an embodiment of the ethos of scholarship.[41] This insecurity was increased by Napoleon: elections were rigged, purges of republican members carried out, police pressure placed on those who resisted, internal financial arrangements altered to the detriment of the corresponding members in the provinces.[42] There was no doubt of the Institut's lack of autonomy in the face of the state. Nor were such changes merely organizational. After the purge in 1803 of the *idéologue* members of the Institut, under the guise of the abolition of its third class, few major works of strongly *idéologue* tendency appeared in France; without political position or intellectual forum, the movement faltered.[43] Institutions in the Napoleonic state cannot be seen as secure bases of power for the individual patron; they too were subject to the political perturbations of Napoleonic France. While the validation of membership of them was essential for the patron on another level, such membership did not ensure immunity from the action of politics.[44]

Our final topic of enquiry must be the effect of this particular type of patronage system on the actual conduct of science.[45] It has often been observed[46] that patronage systems lay stress on fidelity to persons rather than to principle, and inculcate loyalty as one of their strongest ideals. In these systems, therefore, the patron's view of this intellectual field performs many of the functions of institutionally-enforced norms in fully-developed professions. Research programmes were highly personalized; Laplacian physics is a good example.[47] The attack on Laplacian physics which marked the 1820s, therefore, could not distinguish between an attack on a person and on a way of doing science.[48] The patronage system also provided ammunition against itself, through its inability to reconcile, in times of conflict, the ideology of science with the political world which funded and endorsed scientific patrons. The ideology of natural science stressed objectivity of mind as the supreme scientific virtue, and was sure that such objectivity was impossible to attain amidst the distracting and corrupting pressures of human social and political society.[49] Thus that part of a patron's prestige which depended on his being seen as a true natural philosopher could at any point be called into question through his functions as patron in the real political world. An attack in this form could effectively mobilize the rest of the scientific community and the general public, whose idea of what science should be was derived almost entirely from its ideology. This is why, in the classic conflict of this period, that between Cuvier and Geoffroy St Hilaire, public opinion mobilized through newspaper reports, could play such a large part in the affair.[50] It was not possible, once conflict had been brought out into the open,[51] for the contradiction between the ideology of science and its implementation in the political world through the patronage network ever to be resolved. Institutions could be not only the means of validation by the state of the activities of patrons; they could also provide forums for the expression of opposition to such patrons themselves. The ideology of scientific objectivity provided a ready-made plan of campaign. Young outsiders, such as Raspail, could repeatedly point to their moral independence from the corrupt patronage networks which dominated access to the hearing of the scientific community for the beginner in science. In doing so, they claimed higher status for themselves as scientists because of the repudiation of the political context of science, as Raspail did repeatedly in the pages of the *Journal des sciences d'observation*. Refusal of the patronage network became a guarantee of personal authenticity and sincerity, and hence of one's ability to conform to the ideal of the true scientist. To belong to science was to belong to no one else.

Attacks on figures like Laplace and Cuvier are important not only because they were expressions of personal conflict, generational feuds, and impatience with a certain style of political and social control in science; they also threatened the way science itself was done. From the Empire, the production of large-scale synthetic works dominated science and gave it prestige both at home and abroad. The vast scope of such works as Laplace's *Traité de mécanique céleste*, and the *Histoire naturelle des poissons* and the *Règne animal* of Cuvier, provided tasks to parcel out to protégés, both as a reward for loyalty and as a proving ground for competence; they were patronage resources, in fact, as much as was plural office-holding. Biot, for example, earned his right to the protection of Laplace by his reworking of the *Traité de mécanique céleste*;[52] Valenciennes and Latreille sunk their work respectively in the *Histoire naturelle* and in the etymology sections of the *Règne animal*. Such employment was as much an apprenticeship in science as is the modern research laboratory. Conversely the more 'philosophical' style of speculation in natural science represented by Lamarck's *Philosophie zoologique*[53] could not have attracted pupils in significant numbers because it did not generate enough specific tasks to be performed by the protégé under the general direction of the patron. This is one of the reasons that Lamarck found himself on the sidelines of the organized world of science and its connected patronage politics. It also became very easy to down-grade and ignore Lamarck's achievement: able protégés, after all, were useful in keeping the patron before the public eye. They could perform, under his inspiration, successful experiments which would attract admiring reports from the committees of the Institut assigned to their evaluation. Their labours, though duly acknowledged, went to swell the size and number of the patron's books. They could deliver the patron's lecture courses in his name if the number of his appointments prevented him from lecturing himself. Each patron, in competition with other patrons for power and prestige, liked to be able to demonstrate the superiority of his protégés. The simplest and most certain way to do this was through encouraging them to set up experiments which could easily be repeated, to collect facts and to describe new species which could be counted and added to the patron's tally. It is not surprising that the ideal of science favoured by the patrons tended to emphasize the listing of 'positive' knowledge[54] and to denigrate speculative, non-repeatable, non-demonstrable science.

The fact that major projects in science were centred around individual patrons, however, lent them a certain fragility. Attacks on individuals are always easy to make, as has been already mentioned in the case of Laplace, and at this period they were readily turned into attacks on entire programmes in science. After the Second Restoration, pluralism in the holding of office was made illegal in some combinations, and financially penalized in most others. The attack on pluralism was of major importance, because it undermined the power of patrons which under the Empire had been built on the holding of a wide spread of positions, and on the ability of one individual to combine in his person many different forms of power. In this situation men too powerful to attack before 1815 suddenly became more vulnerable. This gradual loss of control by the older scientists, which was not effectively compensated for by the provision of strikingly original scientific programmes by the young, may have been as much responsible for the collapse of French science in face of competition from Germany as was the institutional rigidity of the French system.[55]

In conclusion, this paper has suggested that theories of the institutionalization and professionalization of science are not particularly helpful to us in describing the ways in which scientific institutions were underwritten by the state, or in explaining the complex operation of the patronage system. It has been argued that the relationship of institutions to the state meant that intellectual life could not escape the influence of changes in political life. Science faced peculiarly acute problems in this context because its ideology as a vocation, with its emphasis on the abstraction of the true scientist from the political world, was peculiarly resistant to the justification or support of its political organization. When systems of scientific patronage functioned effectively, as they appear to have done under the Empire, a strong network of caution, gratitude, and reciprocal need prevented this contradiction from rising to the surface. After 1815, the political field changed in such a way as to threaten many of the most characteristic features of the power exercised by successful scientific patrons. The *cumul* was severely restricted, and the freeing of the newspaper press and the growth of a significant public opinion made it easier for dissidents in science to find a wide audience for their attack on the patrons. The expansion of intellectual unemployment and the growth of a self-conscious 'bohemian' intelligentsia also made it likely that job resentments and social philosophy would combine in an attack on the holders of power. In such attacks, the disjunction between the apolitical ideology of science and its life as an organized discipline in the real political world was exploited to the full as a moral justification for intellectual divergence. Such attacks as that contained in the great conflict between Cuvier and Geoffrey St Hilaire, found a forum in the Academy of Sciences itself. Once patronage had ceased to function smoothly, the very openness of the Academy to the influence of general political change meant that it lost the autonomy which would have enabled it to function as a forum for the resolution of disputes between members. As the 1830 revolution approached and political tension heightened, such disputes as that between Arago the man of the Opposition and Cuvier the man of the Government, multiplied and increased in bitterness. Fundamentally it had been a smoothly-working patronage system, rather than institutions, that had stabilized the life of science.

In the last analysis, our major area of ignorance lies around the idea of vocation. It is this idea which links science with one of the most important repertoire of roles in European culture, that of the cultivator of special knowledge.[56] In the case of science in this period, vocational ethos not only gave a plan of campaign against the over-powerful; it also contained the means of intellectual and personal survival in a world where the status of science was often uncertain, and success in the patronage system of science seemed often to demand the loss of personal authenticity in the face of the overwhelmingly personalized power of the patron. Study of vocation would also involve study of the ways in which merit was ascribed within the scientific community.[57] Above all, the disjunction between the ideology of the scientific vocation and the political life of science supplied a perpetual potential for conflict, and hence one possible motor of change amongst others. It is perhaps through approaches such as these that we should examine the achievements and transformations of French science from the Revolution to the July Monarchy.

Notes

1 See, for example, E. Maindron, *L'Académie des sciences,* Paris, 1888; A. Prévost, *L'Ecole de santé de Paris, 1794–1809,* Paris, 1901; idem, *La Faculté de médecine de Paris de 1794 à 1800,* Paris, 1900; Pinet, *Histoire de l'Ecole polytechnique,* Paris, 1887, C. Bresson, *Historie de la chaire d'anatomie à l'Ecole d'Alfort,* Paris, 1928; L. A. Sédillot, *Les professeurs de mathématique et de physique générale au Collège de France,* Rome, 1869; A. Lefranc, *Histoire du Collège de France depuis ses origines jusqu'à la fin du premier Empire,* Paris, 1893. Contemporary institutional histories include J. P. F. Deleuze, *Histoire et description du Muséum royal d'histoire naturelle,* 2 vols., Paris, 1823; J-L. Moreau de la Sarthe, *Mémoire de l'histoire de l'Ecole de médecine de Paris de 1795 à 1822,* Paris, 1824; J. B. Biot, *Essai sur l'histoire générale des sciences pendant la Révolution française,* Paris, 1803. The tradition has survived strongly into this century in, for example, J. Petot, *Histoire de l'administration des ponts-et-chausées, 1599–1815;* G. Chesnau, *L'Ecole des mines,* Paris, 1931; and to a lesser extent in J. Favet, *La Révolution française et la science,* Paris, 1960; E. Ackerknecht, *Medecine at the Paris hospital, 1794–1848,* Baltimore, 1967. Lastly, L. Denise, *Bibliographie historique et iconographique du Jardin des plantes,* Paris, 1903, remains useful for much besides. General histories of education should also be consulted, such as L. Liard, *L'enseignement supérieur en France,* 2 vols., Paris, 1894; F. B. Artz, *The development of technical education in France, 1500–1850,* Cambridge Mass., 1966.
2 Recent studies explicitly treating these problems include M. P. Crosland, 'Development of a professional career in science in France', *Minerva,* 1975, *13,* 38–57, reprinted in idem (ed.), *The emergence of science in western Europe,* London, 1975, pp. 139–60; R. Hahn, *The anatomy of a scientific institution: the Paris Academy of sciences, 1666–1803,* Berkeley & London, 1971. The use of such models was reappraised in R. Hahn, 'Scientific careers in eighteenth-century France', in M. P. Crossland (ed.), op. cit., pp. 127–38; R. Fox, 'Scientific enterprise and the patronage of research in France, 1800–1870', *Minerva,* 1973, *11,* 442–73, reprinted in G. L'E. Turner (ed.), *The patronage of science in the nineteenth century,* Leyden, 1976; J. Ben-David, *The scientist's rôle in society: a comparative study,* Englewood Cliffs, 1971, chapter VIII, an earlier version of which appeared as 'The rise and decline of France as a scientific centre', *Minerva,* 1970, *8,* 161–72. However, these studies still move against the back-drop of professionalization and institutionalization; what is now needed is a double re-examination of the internal life of the scientific community and of its political context.
3 N. Reingold, 'Definitions and speculations: the professionalization of science in America in the nineteenth century', in A. Oleson and S. C. Brown (eds.) *The pursuit of knowledge in the early American republic,* Baltimore & London, 1976, pp. 33–69.
4 C. Diehl, *Americans and German scholarship, 1770–1870,* New Haven & London, 1978, chapter I; R. S. Turner, 'Reformers and professional scholarship in Germany,

1760–1800', in L. Stone (ed.), *The university in society,* 2 vols., Princeton, 1975. Vol. ii: *Europe, Scotland, and the United States from the sixteenth to the eighteenth centuries,* pp. 459–531; H. Dieckmann, 'The concept of knowledge in the *Encyclopédie*', in *Essays in comparative literature,* St Louis, 1961, pp. 73–107; Hahn, op. cit. (2, 'Scientific careers') denies the possibility of a research career in the modern sense in eighteenth-century France.

5 Crosland, op. cit. (2), gives information on the numbers of degrees in science being awarded at the end of this period by the Facultés, as well as on the growth of the numbers of subordinate positions at the Muséum and at the Bureau des Longitudes. But he has still to prove that such positions, treated in his account as part of a *cursus* in 'professional' science, were awarded in relation to success in examination, or that such examinations were regarded as the only or the main way in which candidates for such posts were defined. Medical competence was of course the subject of examination throughout this period, as long before, and medical training was often used by individuals who wished to pursue scientific enquiry while insuring themselves, through medical practice, against its financial risks. But medical knowledge was no necessary guide to competence in the more general field of the life sciences, which remained very largely outside the examination structure of early nineteenth-century France. For the combination of medicine and natural history, see J. Théodoridès, 'Quelques documents inédits sur Toussaint Bastard (1784–1846), médecin et naturaliste', *Histoire des sciences médicales,* 1967, *1,* 1–10.

6 Hahn, op. cit. (2, *Anatomy*), p. x. Of those under discussion, this book is by far the best defined in its use of sociological theory. Elsewhere, vague formulations of key concepts recur; e.g., Ben-David, 'Science was now institutionalized in the sense that scientists . . . could aspire to all the honours and influence they might have wished for', op. cit. (2), p. 89; R. K. Merton, *Social theory and social structure,* Glencoe, 1968, remains the most important compendium of functionalist ideas. In this section I benefit from S. Shapin, 'Origins of scientific institutions' (unpublished typescript).

7 Besides the works listed in nn. 1 and 2, above, unpublished work in progress includes detailed study of the Muséum by Camille Limoges, of the Société philomatique by Jonathan Mandelbaum, of the Bureau des longitudes by John Schuster, of the Lycée républicain by Judith Powell, of provincial learned societies in the nineteenth century by Robert Fox, and of the Académie des sciences in the same period by Maurice Crosland.

8 The best single source of information here remains R. Taton (ed), *Enseignement et diffusion des sciences en France au XVIIIe siècle,* Paris, 1964, supplemented by P. Huard, *Sciences, mèdecine et pharmacie de la Révolution à l'Empire, 1789–1815,* Paris, 1970.

9 Taton, op. cit. (8), pp. 29, 126, 136, 172, 182, 212, 236–9, 293, 357, 390. Deliberately omitted here have been private institutions such as the Lycée de Paris, the scientific courses offered by private individuals, private scientific collections, private scientific societies, the scientific interests of the provincial Académies, and the Académie des sciences in Paris, whose members were mainly elected from among the existing scientific community.

10 L. P. Williams, 'Science, education, and the French revolution', *Isis,* 1953, *44,* 311–30; idem, 'Science, education, and Napoleon I', ibid., 1956, *47,* 369–82. For the increasing pressure of state imperatives on education in the period from 1802, see M. Vaughan and M. Scotford-Archer, *Social conflict and educational change in England and France, 1789–1848,* London, 1971, pp. 16–32, 180–5, 202–30.

11 The *Almanach impérial* for 1813 lists 24 *Académies* or university centres in metropolitan France outside Paris. Sixteen out of this twenty-four had no faculty of science at
all. The eight faculties of science possessed a total of 54 teaching posts (not all of them
filled, and many of them doubling with teaching at the local *lycée*) compared with 237
posts in letters, law, and theology combined. See below n. 12 for the numbers of posts
in Paris.

12 For example, there were three teaching posts at the Jardin du Roi before 1793, when
its new constitution increased the number to twelve – a 400% increase, but still a small
absolute number. The *Almanach impérial* for 1813 gives 19 posts at the Ecole Polytechnique, 6 each at the veterinary schools at Alfort and Lyon, 24 at the Muséum, including 11 *aides-naturalistes,* 8 at the Bureau des Longitudes, 14 at the Paris Faculté de
Médecine, 13 at the Ecole de Pharmacie, 15 at the Faculté des Sciences, 6 at the Ecole
des Mines, and 3 at the Collège de France. The number of posts in general science in
the capital is of crucial importance, since a truly successful career in science was only
possible through admission to the Parisian scientific world. These figures, however, only
gain real significance when measured against a) the demand for the teaching of science,
and b) the numbers of potential aspirants to each post. Numbers for a) could relatively
easily be established from the sources in nn. 1 and 2 above, and more impressionisti-
cally from those in n. 34 below, as well as from any necessary archival research in the
records of scientific teaching institutions, many of which are listed in the bibliography
to Hahn, op. cit. (2, *Anatomy*). For b) the absence of a formal competition system under
the Empire makes calculations more difficult. Perhaps one method would be to compare
numbers of posts with numbers of members of private scientific societies at any given
time, as an indicator of the unabsorbed potential members of scientific institutions.

13 In the Muséum, for example, the Lucas family, directly descended from Buffon,
monopolized lower administrative posts; Frédéric Cuvier directed the menagerie whilst
his elder brother Georges regularly headed the governing body of the Muséum; André
Thouin directed the botanical gardens, whilst his brother Jacques was in charge of the
Muséum's secretariat. See A. Lacroix, 'Une famille de bons serviteurs de l'Académie
des sciences et du Jardin des Plantes: les Lucas', *Bulletin du Muséum national d'histoire
naturelle,* 1938, *10,* 440–71.

14 Use of the *Almanach impérial,* its predecessor the *Almanach national* and its successor the *Almanach royal,* under the headings for the institutions listed in n. 12 above,
gives the following individuals as holding two or more posts within the scientific orbit.
Additional posts held in general administration are specifically indicated.

> 1798–9: Vauquelin, Deyeux, Fourcroy (also député to the Convention nationale),
> Portal, Daubenton.
> 1810: Laugier, Desfontaines, Hallé, Delambre (also conseil général of the Impe-
> rial University), Biot, Poisson, Cuvier (also Imperial University), de Jussieu,
> Portal, Thenard, R-J. Haüy, Geoffroy St Hilaire, Brongniart (also Sèvres por-
> celain manufactory), Huzard, Girard.
> 1813: Vauquelin, Laugier, Hallé, de Jussieu, Corvisart, Portal, Geoffroy St Hi-
> laire, R-J. Haüy, Cuvier (also Conseil d'Etat and Imperial University), Dubois,
> Delambre (also Imperial University), Biot, Poisson, Thenard, Desfontaines.
> 1830: Laugier, Delafosse, Duméril, Clarion, Guilbert, Portal, Desfontaines, Geof-
> froy St Hilaire, Cuvier (also Conseil d'Etat, Minstère des Cultes, Conseil royal
> d'instruction publique), Mirbel (also Conseil royal d'argriculture), Biot, Lac-
> roix, Recamier, Thenard (also député).

Also to be considered are such cases as Berthollet, from 1804 a senator and administrator of the Mint, but holding no formal connexion with a state institution of science, and Lacépède, holding a chair at the Muséum as his only scientific post, yet also directing the Légion d'honneur; and Fourcroy, a member of the Conseil d'Etat since 1805, and director of public instruction from 1804 until 1808 (died 1809).

See also the non-pluralistic science careers of those mentioned in n. 25 below.

The numbers are too small to permit much meaningful statistical juggling; nevertheless, some remarks may be made. From 1810 to 1830, numbers of pluralists, ie., those in a position to exercise really effective patronage, remained fairly constant (13 in 1810, 16 in 1813, 14 in 1830). Total numbers of post-holders in these years are respectively 79, 88, and 89. No striking new concentrations of power can thus be discerned after the Empire (compare the 1798–9 ratio of 6 pluralists to 71 office-holders). It should also be remembered that the Empire saw a continuous expansion of posts, with the foundation of the Faculté des sciences, which makes the steadiness of the ratio all the more striking. Precise forms of pluralism are also interesting; it was rare for office-holders in the 'applied science' institutions also to hold posts in the 'pure science' sector. Posts at Alfort and Lyons, the veterinary schools, and at the Ecole des Mines were rarely if ever held in conjunction with appointments at the Collège de France or the Muséum, for example. There is a virtually complete overlap between membership of the scientific section or first Class of the Institut, and appearance in the list of pluralists. Lastly, it will also be noted that the 'super-pluralists', such as Cuvier, who combined multiple office in both science and general administration were in fact rare; more common were figures such as Lacépède, pluralist only in virtue of one post in each sphere. Pluralism was a valuable patronage resource because it generated work which could be delegated to protégés, and gave them income.

15 See n. 5 above.

16 For the fierce running battle between the Imperial University and the army, for example, see D. Outram, 'Education and politics in Piedmont, 1796–1814', *Historical journal,* 1976, *19,* 611–33.

17 The administration of the Imperial University was controlled at least as much by literary men as it was by scientists. See Outram, op. cit. (16), p. 628. Biographies of such literary figures include A. Wilson, *Fontanes: essai biographique et littéraire, 1757–1821,* Paris, 1928; P. de Reynal, *Les correspondants de Joubert,* Paris, 1883; A. V. Arnault, *Souvenirs d'un sexagénaire,* 4 vols., Paris, 1835, iii, 292. On the general topic of the influence of this 'literary group' on the evolution of educational policy, see D. Outram, *Education and the state in the Italian départements annexed to France, 1796–1814,* University of Cambridge PhD thesis, 1974, chapter IV.

18 Forthcoming work by Barbara Haines on St Simon and by John Hooper on Fourier will help to clarify some of these points. Still valuable is F. A. Hayek, 'The counter-revolution of the engineers' *Economica,* 1941, *8,* 119–150, 281–320. The implications of this growing division between pure and applied science cannot be treated at length here; but see n. 14 above and n. 23 below, for some of the issues raised.

19 For a general approach to this problem, see L. O'Boyle, 'The problem of an excess of educated men in Western Europe, 1800–1850', *Journal of modern history,* 1970, *4,* 471–95.

20 D. Roche, 'Milieux académiques provinciaux et société des lumières', in G. Bollème, J. Ehrard, F. Furet, D. Roche, and J. Roger (eds.), *Livre et société dans la France du XVIIIe siècle,* Paris, 1975, pp. 93–184.

21 This conflict between 'clean' power from knowledge of nature, and 'dirty' power from the political world, has been examined in D. Outram, 'The language of natural power: the funeral *éloges* of Georges Cuvier', *History of science,* 1978, *16,* 153–78.

22 E. Brambilla, 'Professioni giuridiche e mobilità sociale nella Francia pre-rivoluzionaria', *Studi storici,* 1978, *4,* 819–30.

23 See D. Roche, 'Talents, raison et sacrifice: l'image du médecin des lumières d'après les éloges de la Société royale de médecine, 1776–1789', *Annales: économies societés civilisations,* 1977, *32,* 86686. See also G. Cuvier, *Eloges historiques,* 3 vol., Paris, 1819–27, i, 38, 167; Outram, op. cit. (21); the theme of the break with the family and with human society in general as a condition of the attainment of certain forms of knowledge, is of course very firmly rooted in Christian culture. See L. Theis, 'Saints sans famille? Quelques remarques sur la famille dans le monde franc à travers les sources hagiographiques', *Revue historique,* 1976, *155,* 3–20. Only the private forms of scientific community seem to have been able to develop a truly cooperative ethos. It was only in the good fellowship in natural enquiry insisted upon, for example, by the Société philomatique, that the two conflicting definitions of prestige – one asocial because scientific, and the other political because institutional and state-guaranteed – were able to be resolved. I rely here on unpublished work by Jonathan Mandelbaum.

24 G. Chaussinand-Nogaret, L. Bergeron, and R. Forster, 'Les notables du Grand-Empire en 1810', *Annales: économics, societés, civilisations,* 1971, *26,* 1052–75. Many scientists such as Berthollet and Cuvier were also awarded state endorsement through gifts of land and revenues in the occupied territories of the Empire. See M. Senkowska-Gluck, 'Les donataires de Napoléon', *Revue d'histoire moderne et contemporaine,* 1970, *17,* 680–93. Lacépède's direction of the Légion d'honneur implicated science in the honorific system of the Empire even further. Numbers of scientists given titles and other honours by Napoleon and the Bourbons might easily be calculated from the *Almanach imperial* and *Almanach royal.*

25 For Napoleon's personal interest in the Institut, see G. Lacour-Gayet, *Bonaparte membre de l'Institut,* Paris, 1923. Chaptal, for example, became minister of the interior from 1800 to 1804, as Laplace had for six disastrous weeks in 1799. Fourcroy's political career is well known. For other such examples see n. 14 above. On Fourcroy himself see G. Kersaint, 'Antoine-François de Fourcroy, 1755–1809', *Mémoires du Muséum national d'histoire naturelle,* série D, 1966, *1,* 1–296. It is noticeable, however, that with the settling of the political arena after the 1803 purge of remaining republicans, it became uncommon to find intellectuals occupying posts as elected representatives. Increasing property qualifications for election also intensified this development. The same is true of the Restoration, except, for obvious reasons, for 1815 and 1830.

26 Biot, op. cit. (1), p. 1.

27 Hahn, op. cit. (2, *Anatomy*), p. 262, argues that bureaucratic employment of scientists in the early years of the Revolution caused a traumatic abandonment of the scientific for the bureaucratic ethos. He does not, however, prove the previous existence of this latter ethos. The confusion of the years of the Terror, and the intense personalization of power under the Consulate and Empire would seem rather to indicate that if anything, the debate within science over the nature of power was not altered; see n. 20 above. Recognisable ideals of bureaucratic government did of course exist in eighteenth-century Germany, but their influence in France is unclear, and even in Germany itself severe confusion was caused by their subjection to the arbitrary personal power of the ruler. See M. Raeff, 'The well-ordered police state', *American*

historical review, 1975, *80,* 1221–43. The growing influence of biological metaphors on social description is an example, rather, of overlap between science and field of government.

28 R. Fox, 'The rise and fall of Laplacian physics,' *Historical studies in the physical sciences,* 1974, *4,* 89–136.

29 E. Frankel, 'Career-making in post-revolutionary France: the case of J. B. Biot', *British journal for the history of science,* 1978, *11,* 36–48. Formal competition for posts under the Empire was used far less than a system of recommendation to the competent Minister; this of course reinforced the power of the patron.

30 Geoffroy St Hilaire was instrumental in saving both Haüy and Daubenton from imprisonment and persecution under the Terror. Such actions forged indissoluble links with important patron figures. If the Terror shattered the unity of the scientific community as it had existed in 1793, it also provided opportunities for young recruits to lay immense purchase upon it. See I. Geoffroy St Hilaire, *Vie, travaux et doctrine scientifique d'Etienne Geoffroy St Hilaire,* Paris, 1847, chapter 1.

31 I owe this definition of patronage efficiency to Jean-Claude Guédon and Camille Limoges.

32 For a review of this literature, see E. Gellner, 'Patrons and clients', in E. Gellner and J. Waterbury (eds.), *Patrons and clients in Mediterranean societies,* London, 1977, pp. 1–6. J. Ben-David, op. cit. (2), p. 3, nn. 3–4, gives a comprehensive review of sociological literature concerned with modern science, none of which addresses itself specifically to this problem in a historical context of a Western preliberal state.

33 Joseph-Marie Degérando (1772–1842) could be taken as representative of this subclass of patron. Though not himself scientifically active, his contributions to the *Décade philosophique,* his membership of the Institut, his important administrative posts in the Ministry of the Interior and in the administration of Genoa and Tuscany under Napoleon, gave him a range of contacts in the whole politico-cultural world of Europe which few could have equalled, and which resulted from his mixing of many distinct sources of power. Again, Cuvier's early patrons included not only such obvious figures as Lacépède, but also the politician Miot de Mélito; see Cuvier's 'Autobiography', MS Flourens 2598 (3) of the library of the Institut de France.

34 Comprehensively listed in J. Tulard, *Bibliographie critique des mémoires sur le Consulat et l'Empire écrits ou traduits en français,* Geneva & Paris, 1971.

35 See E. Hatin, *Bibliographie historique et critique de la presse périodique française,* Paris, 1866, reprinted Turin, 1960 – an invaluable tool here. For some of the societies, see M. Agulhon, *Le cercle dans la France bourgeoise, 1810–48: étude d'une mutation de sensibilité,* Paris, 1977. Study of the resulting lists of names and the different structures of social and cultural affiliation would reveal much about the structure of contacts within the politico-cultural élite.

36 D. Bryant, 'Revolution and introspection: the appearance of the private diary in France', *European studies review,* 1978, *8,* 259–72; D. Outram, op. cit. (1) lists autobiographies produced by members of the scientific section of the Institut to be used in their funeral *éloges.*

37 The best recent study of general applicability of the *salon* as a problem in the sociology of knowledge is D. Hertz, '*Salonières* and literary women in late eighteenth-century Berlin', *New German critique,* 1978, *1,* 97–108. The problem of attracting the attention of a Parisian patron was of course intensified if the aspirant came from a provincial milieu. In this context, correct handling of the letter form assumed great importance.

See some of Cuvier's first letters in this direction to Oliver, in J. Théodoridès, 'Jean-Guillaume Bruguière 1749–1798, et Guillaume-Antoine Oliver, 1756–1814, médicins, naturalistes et voyageurs', *86e Congrés des sociétés Savantes*, 1961, pp. 173–83.

38 Social anthropology may also be utilized in these contexts; e.g., R. Paine, 'What is gossip about?: an alternative hypothesis', *Man*, 1967, *2*, 278–85.

39 See for example Laplace's graceful bowing-out speech to Biot, quoted in Frankel, op. cit. (29). This whole article is a repertoire of such *gestes* between patron and client.

40 Gellner, op. cit. (32), p. 3: '[as such] . . . the liberal state . . . does not give rise to patronage relations'.

41 Hahn, op. cit. (2, *Anatomy*), pp. 311–12, emphasizes the fossil, static character of the Institut as the embodiment of the values of the state, rather than as a dynamic forum for the production of science.

42 *Mémoires de Larévellière-Lépeaux, membre du directoire exécutif de la république française et de l'Institut national, publiés par son fils*, 2 vols., Paris 1895, ii, 463–81.

43 *Idéologie* is presented as surviving as a diffused tendency in French culture in S. Moravia. *Il tramonoto dell illuminismo: filosofia e politica nella società francese, 1770–1810*, Bari, 1968.

44 See the description of the attacks on Laplace, for example, in R. Fox, op. cit. (28).

45 Extended discussion of the nature of 'scientific method' in this period would cover too much space to be attempted here. Such questions as the nature of objectivity and of 'positive knowledge' will be treated largely as they were utilized in the power-structure of organized science. The question of how far such a methodology was *produced* by this organization must be left for more extended treatment than this article can provide.

46 Gellner, op. cit. (32), pp. 1–2.

47 Fox, op. cit. (28).

48 See n. 44 above for Laplace; a representative attack on Cuvier on these grounds is contained in [A. B. Eymery], *Dictionnaire des girouettes; ou, nos contemporains peints d'après eux-mêmes*, 2nd edn., Paris, 1815: 'Cuvier'. Cuvier's former protégé Henri Ducrotay de Blainville was to base his entire analysis of the faults in Cuvier's style of science on what he conceived to be the fundamental flaws in Cuvier's character; see his *Cuvier et Geoffroy St Hilaire: biographies scientifiques*, Paris, 1890. It should be noted that although M. P. Crosland's *The society of Arcueil: a view of French science at the time of Napoleon I*, London, 1967, is avowedly a study of patronage, it seeks to abstract this patronage from the political field, and to maintain that '. . . the familiar political and military history must be put to one side . . . Men of science tend to be less sensitive to political changes than scholars in other fields . . . ', (p. ix).

49 D. Outram, op. cit. (21).

50 The most thorough single study of this conflict is T. A. Appel, *The Cuvier-Geoffroy debate and the structure of nineteenth-century French zoology*, University of Princeton PhD dissertation, 1975.

51 Cuvier's reasons for repressing public discussion of his scientific differences with Lamarck were concerned with the maintenance of a respectable public 'image' for science; they thus also illustrate the extent to which science did not have a secure place as a vocation in French life. See R. W. Burkhardt Jr., 'J-B Lamarck, evolution, and the politics of science,' *Journal of the history of biology*, 1970, *3*, 275–98.

52 Frankel, op. cit. (29). Cuvier's *Règne animal* appeared in 4 vols., Paris, 1817, and in a second edition in 5 vols. in 1829; vols. i and ii of the *Histoire naturelle des poissons* appeared in 1828, iii and iv in 1829, v and vi in 1830, and vii and viii in 1831. After

Cuvier's death in 1832, the work was carried to completion by Valenciennes alone. See also Pierre-Simon Laplace, *Traité de mécanique céleste*, 5 vols., Paris, 1799–1825.

53 *Philosophie zoologique, ou exposition des considérations relatives à l'histoire naturelle des animaux, à la diversité de leur organisation et des facultés qu'ils en obtiennent, aux causes physiques qui maintiennent en eux la vie et donnent lieu aux mouvements qu'ils executent; enfin, à celles qui produisent les unes les sentiments et les autres l'intelligence de ceux qui en sont doués,* 2 vols., Paris, 1809.

54 This for example is the ideal of science explicitly favoured in the preface to the 1817 edition of Cuvier's *Règne animal*.

55 For development of this theme, which is again too large to be properly treated within this paper, see for example Ben-David, op. cit. (2, *Minerva*), R. Fox, op. cit. (2).

56 Withdrawal from society, for example, was early identified as a condition of the attainment of insight and objectivity. See P. R. L. Brown, 'The rise and function of the Holy Man in late antiquity', *Journal of Roman studies,* 1971, *61,* 80–101. The end of the eighteenth century saw as much confusion over the signs by which an authentic *savant* should be recognized as it did over the definition of élite status in general, as can be seen in the debates chronicled in R. Darnton, *Mesmerism and the end of the Enlightenment in France,* Cambridge, Mass., 1968. The typologies of the funeral *éloges* (see n. 21 above) may also represent attempts to solve these problems of definition.

57 An example of the application of models of vocation to scientific careers is R. Porter, 'Gentlemen and geology: the emergence of a scientific career, 1660–1920', *Historical journal,* 1978, *21,* 809–36.

4

THE ORDEAL OF VOCATION

The Paris Academy of Sciences and the Terror, 1793–95

From: *History of Science*, 21 (1983), 251–273. © Sage Publications.

Brief though it was, for many of its contemporaries the period of the Jacobin Terror remained an undoubted epoch of division.[1] As political change violently accelerated, new roles, new opportunities, new tests, new dangers, swung into the forefront of men's minds, cut them off from past experience, and propelled them towards a compelling, threatening, violent, and deeply moving vision of the future. The Republic of Virtue, it seemed, could be achieved only at the cost of that river of blood, which as Chateaubriand reminded his readers in a famous passage "separates for ever the old world in which you were born from the new world on whose frontier you will die".[2] Yet in spite of the deep scar which the Terror left on the consciousness of contemporaries, we still lack a definitive study of its effects on French science, as on many other aspects of French society. This paper should be regarded as the first step towards such a study. Attention has been confined to those men who were members of the Royal Academy of Sciences at the time of its abolition on 8 August 1793.[3] There is no intention to claim that these groups were in any way identical with the whole community of science, whatever meaning should be attached to that term. Nor is there the intention of re-examining the theses of 'Jacobin science' advanced by Gillispie, Williams and Hahn.[4] If we may take it that the Jacobin hostility to men of science was caused by dislike of a certain form of science, it would on the other hand certainly be unsafe to explain the fate of all individuals solely by reference to this model. The concern here will be primarily with the effects of the Terror period, for whatever reason, on individual men of science.

Our first step must be to establish how the Terror affected the individuals in our group. Once this prosopography of the Terror has been established, the abundance of autobiographical writing produced in this period allows us to mount a more detailed study of the reactions of individuals to these experiences, and of their effect on their sense of vocation as men of science. We will then turn to a consideration of the complex historiography surrounding the fate of French science, and the Academy of Sciences in particular. This historiography, sometimes approaching the status of myth, began to be created in the weeks immediately following the fall of Robespierre, and was deeply influenced by contemporary and later discussions on the meaning of the Terror and the whole Jacobin era. In the late nineteenth and twentieth centuries, these debates have become an integral part of the conflict between Left and Right in France for control of the interpretation of the history of the revolutionary period, and hence of the interpretation of the nature of the French national experience. These political inputs need to be more explicitly discussed than is often the case because they still influence accounts of this period in ways which are not fully allowed for by historians of science. Finally our discussion will move forward into an evaluation of the social and psychological effects of the Jacobin Terror on the men who formed the First Class of the Institut from its foundation in 1795, and will focus on Roger Hahn's account of this same period.[5]

The organization of the regime of the Jacobin Terror, in response to the national emergency of 1793, has often been described, and there is no need to enter into a detailed account here.[6] More germane to our purposes is an assessment of how the regime of the Terror altered life for the individuals with whom we are concerned. The changes of the Terror appeared most noticeably in the passing by the Convention of the Law of Suspects on 17 September 1793, which legalized the imprisonment of all persons whose loyalty to the revolution and to the Republic was held to be doubtful. The definition of doubtful loyalty lay in practice with district Revolutionary Committees, bodies which tended to be composed of convinced Jacobins drawn from the lower-middle and working class. The Law of Suspects thus left the way open for the settling of personal and class hatreds which had little to do with attitudes to the Revolution. Denunciations and counter-denunciations thus became commonplace; family loyalties, class boundaries, professional prestige, friendship, patronage, no longer acted as any defence. The individual was left at the mercy of his neighbours, subjected to an intense scrutiny by them and by the state which was unparalleled in the experience of most men, and meant that the slightest word or gesture could, if misinterpreted, end in denunciation, imprisonment and death.[7] The Law of Suspects also increased the power of the Revolutionary Tribunal in Paris, the central judicial body involved in the prosecution of crimes of opinion, with suspected traitors and counter-revolutionaries. Paris henceforth became not only an expensive and hungry place to be, with the *crise de subsistances* of 1793–95, but also a dangerous one. The machinery of inspection of individual opinion was pervasive, and the means of repression close to hand.

All these developments also lent a new importance to the Committee of Public Safety, set up in April 1793 on the initiative of the Convention, with the aim of supervising and invigorating the executive. After 5 September 1793, the Committee for all practical purposes took over the business of government. It was to continue in existence even after the fall of Robespierre and be the agent of the anti-Jacobin purges carried out by the Directory; it was dismantled only in 1795, as was the Revolutionary Tribunal itself. The Committee centralized the organization of the nation for war. A mass conscript army was raised for the first time in French history; this, the enforcement of controls on the food supply, as well as the continual interrogation of individual political loyalties, meant that the Jacobin state exerted an unprecedented degree of control over the lives of individuals. At the same time, the most elementary safeguards for accused persons brought before the Revolutionary Tribunal completely collapsed. The law of 22 *prairial* (10 June 1794) deprived accused persons of any rights of defence. Appearance before the Tribunal became equivalent to a death sentence. As the Jacobin period progressed, the Terror became increasingly severe. Between March 1793 and August 1794, 16,594 persons were executed, and half a million imprisoned.[8] The rate of executions in Paris reached its peak in June 1794. It is interesting to note, however, that the pace of the Terror was not determined by the pace of military events. The Terror heightened at precisely the period when the French armies were already breaking out of the encirclement of 1793. This was because Terror was not only the means which ensured organization for victory; it was also the means by which the Jacobins destroyed their political opponents such as the Dantonists and the Hébertists, within the Convention itself. Not least, of course, Terror was also seen as the perfect tool of social engineering which would produce that equality of all men before the state which the Jacobins regarded as an essential stage in the establishment of the Republic of Virtue.

The Terror ended with the overthrow of Robespierre and his closest associates on 9–10 *thermidor* (27–28 July 1794). But in spite of the fact that the end of the Terror can be dated so precisely, it is noticeable that both contemporaries and modern commentators have been unable to agree on its length. The chemist Fourcroy thought that it lasted fourteen months; the astronomer Lalande, nine; Benjamin Constant allotted it the thirteen months between June 1793 and *thermidor*. The geologist Dolomieu was possessed by an "intimate conviction of terror" as early as 1792; not surprisingly, since he had witnessed the murder of his intimate friend La Rochefoucauld in that year.[9] The only convincing explanation for these discrepancies is that the Terror existed as an attitude of mind long before its institutional and legal mechanisms were created. The collapse of the social and financial structure of the Ancien Régime, and the rapid succession of governments after 1789, together with the conscious use of terror as an instrument of political and social engineering after 1793, meant that no man's future was secure. The word Terror, in this sense, may have been a neologism in 1793, but as a state of mind it already existed.[10] This is a point to which we will return when discussing the psychological effects of the period of the Terror on specific individuals. But now we should turn, first of all, to the effects of the Terror on the Academy of Sciences as a whole.

II

As yet, a prosopography of the Terror as it affected the Academy of Sciences does not exist. Without it, we have no way of estimating the extent to which contemporary interpretations of this period can be accepted as being based in fact, or should rather be interpreted as mythologies useful in political negotiation.

For this section, biographical information was located for as many members as possible both of the old Academy of Sciences and of the Institut.[11] This information, tabulated in Tables I and IV, allows us to draw some conclusions about the impact of the Terror on the men included in these groups. They make it clear that the impact of the Terror was very great. Total mortality in this period ran to one-quarter of the membership over this three-year period. Many deaths not directly due to the Terror must have been hastened by it. The toll of twelve men represents a loss of one-quarter of the members in this class after the reform of the Academy in 1785. Of this number, four deaths directly due to the Terror took place among the forty working Academicians and constitute one-fifth of the total number of deaths. Table I thus makes it clear that the First Class of the Institut could never have been a simple reproduction of the Academy of Sciences: too many had died in the interim.

Table II shows that the Terror also weighed heavily on those who remained alive. Flight or imprisonment were common experiences. Too many individuals were removed from circulation for research programmes or networks of patronage to be effectively maintained. This breakdown of scientific life was enhanced by the effects of the revolutionary wars, which led to the calling up of many doctors and pharmacists into the army and navy. Voluntary withdrawals unconnected with military service also began in 1792, well before the closure of the Academy in 1793, but contemporaneous with the period of the "conviction of terror" experienced by Dolomieu. They clearly have little relation with the order of 17 April 1794 which banned the former nobility from residence in Paris. Imprisonment, however, was almost confined to 1793 itself, in our sample. This is hardly surprising, since it was that year which saw the greatest extension of the powers of the Revolutionary Tribunal. The fact that the Committee of Public Safety also placed all government Commissions under its direct inspection in that year, accounts for the increasing vulnerability of those men of science serving on such bodies as the Commission on metrication. The frequent purges of their members on grounds of suspected royalism are thus not surprising.[12]

Taken as a whole, these tables show that over one-third of those who were to be members of the new Napoleonic Institut suffered some form of violent interruption in their lives. Their numbers were far greater than that of the men who benefited by the war emergency of 1793 to find paid government work. We are thus now in a position to say that contemporary claims that the Academy of Sciences had lost half its membership under the Terror no longer seem the exaggerated product of the myth of 'Jacobin vandalism'.[13] Statements by modern historians that the Terror created no real hiatus in the community of science, now also have to be treated with reserve.[14]

III

When we come to examine the effects of the Terror, therefore, we are not talking about experiences which affected only a few individuals, but about experiences common to the group with which we are concerned, and which were embodied in a remarkable outburst of autobiographical writing. It is obvious how the insecurity of life in a period such as the Terror could make men need to sum up their lives and achievements. But there was more to autobiography than this. It became the main medium through which men struggled to describe the extraordinary experiences they had undergone, and their effects upon their perceptions of themselves. In many cases, the period of the Terror produced the need to re-examine a previously taken-for-granted scientific vocation, to use its rhetoric at a higher power, to provide a sheet-anchor for personalities on the point of disintegration under the stress of flight, imprisonment, solitude and hunger.

The period of the Terror faced men with a difficult adjustment to the new public world of mass politics, party conflict, and a heavily propagandized public sentiment. It also induced a new atomization, which was the paradoxical corollary of the invention of mass politics. As social and emotional bonds between individuals broke down, men became for the first time truly equal, equal in their helplessness before the machinery of the Republic of Virtue. Dolomieu, for example, carried away from the Paris of 1792

> . . . such an impression of fear that I have been unable to rid myself of it. Worry and distrust were printed on every face; no-one drew attention to themselves, no-one dared to venture an opinion, everyone looked out for themselves. Friendship and social intercourse no longer exist; there is nothing but factions and parties.[15]

Fourcroy, not one of the least defenceless actors of this period, a member of the Convention, receiving payment for his work on the production of gunpowder, later recalled that

> I went without bread for ten months. I and the five persons with me lived on potatoes. . . . I hardly saw anyone; I refused invitations, dinners, I was as usual outside every faction.[16]

Sociability in fact had become the dangerous gateway into the world of the political parties; the fate of the Dantonists and the Hébertists showed its risks to the full.

Against these pressures, individuals opposed another sort of rigidity: that of the Stoic ethos of self-sufficiency, an ethos which allowed a man to present an unvarying face to a harsh world, whilst yet cultivating a full knowledge of and contact with himself. This was an ethos whose defensive capacities were widely utilized in this period. It fitted especially well with the existing ideology of scientific work: solitary, asocial, its quality guaranteed by the very autonomy and self-sufficiency of its practitioners.[17]

Possibly the most complete account of such a response is to be found in the autobiography of the botanist, geologist, and future Napoleonic Prefect, Ramond. Imprisoned at Tarbès as a supporter of the Girondins, Ramond's fate lay in the balance for many months:

> In that long solitude, in that deep silence, awaiting the scaffold from one moment to the next, what do you think I did? Think of myself? No, I forgot myself. Thus I left prison without rancour, indeed without anything, dressed in old rags, walking without shoes, living on the few sous that had not been stolen from me; I turned my resentment against events and not against men, I conquered fate by patience, turned for consolation to the study of nature, and always looking to the future, never to the past, I reconquered my position in society by work and economy. . . . I could write you a fine book on the miracles of patience, of work and of economy that I have performed, like some new Robinson Crusoe, with the pride of a true man, who looks to himself for his own security, and is sufficient unto himself.[18]

Ramond was clear that he had to accept an absolute break between his own present and his own past if he was to retain the personal autonomy which he saw as the condition of sanity. He turned to the external objective world of nature to teach him that self-sufficiency by which he survived his imprisonment. The study of nature offered a vision of an outer world which was stable and unthreatening. It also offered the ability to oppose to the terrifying atomism of the Terror a constructive solitude which enlivened and authenticated the individual instead of reducing him to equality with the helpless citizens of the Republic of Virtue. This is not of course to deny that a strong ideology of science as a vocation existed long before the Terror; it is rather to argue that for many individuals the Terror intensified the importance of this vocation to the point where it carried the whole weight of the integrity of the personality. For the whole of the rest of his life, Ramond "always kept back a part of my thinking which was reserved for science".[19] This was not an attitude adopted by a successful Prefect out of scientific professionalization, but out of a psychological necessity first laid upon him in the prisons of the Terror.

Ramond's response to his extreme situation was echoed by others who were not under such a direct threat. The autobiography of the naturalist Bosc provides another case in point. Having fled to Radegonde after the fall of the Girondins, with whom he was closely linked, Bosc changed his way of life towards a more intimate contact with nature:

> I wore working class clothes, worked on the land in the woods, did my own cooking. . . . I spent other days on natural history, on manual labour, and in hunting. Unable to obtain the books I lacked, I undertook a description, with drawings, of forest spiders, as the least well-known.[20]

The experiences of the naturalist Lacepède were similar. Having fled from Paris to Leuville after his royalism attracted unfavourable notice in the capital, he went on working on natural history, buoyed up by an intimate contact with nature:

> Sitting on the ruins which surround the high tower of Montlhéry, looking onto an immense prospect . . . or lying on grass sprinkled with flowers, in the shade of rustling poplars, on the edge of the great lake at Marcoussis, or walking under the green vaults of the vast and solitary forests near the lake, I loved to think of the wonderful effects of the power of nature, and of its sublime laws. . . . I forgot the world, and saw only the universe.[21]

A clear view of nature, in other words, was guaranteed by a retreat into a Stoic *otium*. Retreat into the world of nature remained a necessity for Lacepède long after 1794. Henceforward, he was to resent the interruption of his work in natural history by the administrative duties of his post as first Chancellor of the Légion d'Honneur as "almost unbelievably painful, and to which I have never been able to accustom myself".[22]

Solitude, retreat and *otium* in fact became of more than personal significance under the Terror. In a nation geared up to conquer or die, the retreat into solitude, a pyschological necessity for the individual, became a crime towards the state. The uninvolved man, the self-sufficient man, the solitary man, the man of contemplation represented roles which the state could not endorse. When St Just arraigned Danton before the Convention, he cried: "In stormy debates, we were outraged by your absence and your silence: you, you were talking of the countryside, of the delights of solitude and leisure."[23] That reconciliation between *otium* and *negotium* which Daniel Roche has seen as one of the hallmarks of the pre-revolutionary provincial Academies, was henceforward denied to the *savant*, as it was to anyone else.[24]

The Terror in fact tended to replace the passive role of the man of *otium* with other and more dramatic models. The most obvious model for the imprisoned *savant* was that of Socrates, a great philosopher unjustly persecuted, who continued to teach in prison until forced to commit suicide. Even before the Revolution, David and Canova had contributed significantly to the popularization of images related to Socrates.[25] Comparisons with Socrates became almost ritual in the period of the Terror itself: Lalande spoke of Bailly's fate as resembling that of the Greek philosopher, and Millin de Grandmaison recalled the deaths of his fellow prisoners Roucher, Chénier, Biran, Trudaine and D'Ormesson, as having been supported in a like manner.[26] This was not merely a literary trope, for men did actually act out Socratic roles. Saron went on with astronomical calculations in his condemned cell, Millin continued to write his book on natural history. Cousin gave geometry lessons to his fellow prisoners, Destutt de Tracy meditated on the unity of the sciences.[27]

These men found themselves in a dramatic period of history in the most literal sense. Everyone played roles because of the intense scrutiny to which they were subjected by the state and by hostile neighbours. For the *savants*, the drama of the Terror made manifest the inner dramas of scientific vocation. Often, this strengthened vocational commitments, by exemplifying the struggle against the odds for dedication and objectivity which had always been at the heart of scientific vocational ideology. The theatricality and role-playing of the Terror also had another effect, however; this was to make the period immediately after *thermidor* into one glutted with heroic images all too easily appropriated for the manufacture of myth.

IV

All too often, historians of science have allowed this myth to pass as historiography or even as history. It is now time to subject the mythology of the Terror as it affected the sciences to a more searching, if necessarily brief, examination. Contemporary comment about the Terror seems to fall into three main strands. First comes the thesis of 'Jacobin vandalism', produced in the immediate aftermath of *thermidor* by, among others, the chemist Fourcroy, and the constitutional Bishop of Blois and *conventionnel*, Henri Grégoire. They argued that the Jacobin government had been unparalleled in its destruction of art, architecture and libraries, as well as in its persecution of the *savants*. The second strand, the favourite of Socialist and Marxist historians, such as Albert Mathiez, since the closing years of the nineteenth century, emphasizes, as did the Committee of Public Safety itself, the importance of the nation's organization for war, and sees 'the sciences' as a vital part of that organization. Such historians saw in the government of Robespierre the first national government of France which was also a government of social justice. As the era of the Great War intensified national jealousies, Socialism had to be dissociated from its former internationalism if it was to appear a real political option for France. The rehabilitation of the Jacobins and of the Terror was an attempt to establish an acceptable historical model for such a government. Accordingly, instead of stressing the devastation caused to science, they view the collaboration of a few scientists with the war effort of the Republic as evidence of the integration of science as a whole with the foundation of the modern French nation.

The third strand goes back to the discussions of the effects of the Terror which began immediately after 1794, and which were ably summarized in 1797 by Benjamin Constant.[28] He refused to allow that the Terror had anything to do with creating the unity of France or even with ensuring her military survival. For modern historians who follow this line, such as Georges Bouchard and Joseph Fayet, the heroic *savants* of the Mathiez interpretation become reduced to participants in a horrific deviation from the true course of national development, a deviation which threatened to destroy all the genuine liberal gains of the Revolution.

The myth of Jacobin vandalism began to be produced immediately after *thermidor*. Fourcroy's first speech on 'vandalism' was delivered to the Convention of 31 August 1794. He was quickly followed by Grégoire, who produced three lengthy reports on cultural 'vandalism' by the end of the year.[29] Grégoire claimed that the Committee of Public Safety had produced an organized system for the persecution of men of talent. He supported this claim with lists of those imprisoned and guillotined.[30] These views were echoed in the speeches which opened the first session of the Institut eighteen months later. Many of Grégoire's charges unjustly ignored the work of the Convention in higher scientific education, and in the reorganization of such institutions as the Paris Museum of Natural History.[31] But in relation to the persecution of the *savants*, Grégoire, at least for the sciences, was referring to a reality, as our previous surveys have shown.

At the same time, Grégoire performed interesting operations on other aspects of the truth. He implied, more by *omissio veri* than by *suggestio falsi*, that the Terror had affected only the *savants*. Most contemporaries found that on the contrary the Terror was terrifying precisely because it was indiscriminate, and thus encouraged every man to denounce his neighbour or his professional colleague.[32] Grégoire's version of the incidence of the Terror was concerned to award to the *savants* the morally enviable role of victims, and to obscure the internal divisions which had marked the scientific community under the Terror.[33]

The mathematician Biot was the first to state the line later taken up by Mathiez. In his 1803 study of the history of the sciences under the Terror, he certainly did not make light, as later historians were to do, of the sufferings of individuals. But he did insist that the Jacobin war effort, supported as it was by Terror, had marked a turning point in the public status of the sciences: "Revolutionary despotism gave them a political existence, used them to inspire confidence in the people, to prepare victory and win battles."[34] This is a view with a long legacy for the future. It was not only repeated by French historians of the years of the First World War, but reappears in the work of Taton and even, very recently, in a modified form, in the work of Hahn.[35] It led them to attach enormous importance to the mobilization of the scientists for the Jacobin war effort itself in the formation of the modern French nation.

For Benjamin Constant, on the other hand, it was self-evident that it was the Terror itself which had caused many of France's worst problems.[36] Terror certainly had the effect of causing a public reaction in favour of the monarchy: the frenzy of 1794 had made many men abjure the achievements of 1789. Liberal opinion therefore tended to see the repudiation of the Terror as the only way to stem the tide of counter-revolution, and safeguard the moderate gains of 1789–91.[37] Historians influenced by Constant's views have tended to play down the involvement of science with Jacobins' war effort. More sympathetic than Mathiez to the sufferings of those who did not wish to fit in with the Republic of Virtue, historians like Georges Bouchard have pointed out that the numbers of those scientists who were employed in the war effort were tiny in relation to the numbers of those whose lives were devastated by it.[38] The major beneficiaries – Monge, Berthollet, Chaptal, Fourcroy, Guyton de Morveau, Vauquelin, Vandermonde, Pelletier and Darcet – gathered round them their protégés Hassenfratz, Adet, Desroizilles, Champy, Leblanc and Conté. Chaptal was politically protected by Fourcroy. Hassenfratz and Adet had already worked with Fourcroy under Lavoisier. Hassenfratz had politically protected Vandermonde.[39] What we are seeing here is not the 'mobilization of the *savants*' so beloved of the Mathiez school, but the reproduction of existing patronage networks.

All this examination of historiography, however, fails to answer the question of why contemporaries themselves should have constructed so many myths around the period of the Jacobin Terror. We are not simply confronted with general interpretations of the period containing strong mythological elements, we also have at our disposal many tales concerning individual actions and fates. Albert Mathiez and his followers launched a concerted attack upon these myths relating to individuals, almost all of which reflected badly upon the Jacobin treatment of individual *savants*. Textual criticism and historical verisimilitude exposed story after story as mythological constructions. Bailly's dignified last words were shown to be drawn from far older literary models.[40] Daubenton, far from running risks as a faint-hearted republican, was shown to be thoroughly at ease with his local revolutionary committee.[41] Even the notorious Jacobin slogan, "La république n'a pas besoin des savants", was shown to have been a fabrication by Henri Grégoire.[42] To complete the picture, Mathiez engaged directly with the alarming figure of Robespierre himself, and argued that the Terror, far from being an integral part of the Republic of Virtue, as contemporaries might have supposed, had been a policy forced upon an unwilling leader by the pressure of events.[43]

Mathiez and his followers performed the service of drawing attention to the ways in which such legends were constructed and diffused. He did not, however, enquire into the functions which such myths performed for contemporaries – other than to attack the Jacobin Republic – and thus failed to provide an explanation for their existence in the first place. Their *raison d'être* was in fact far more complex than anti-Jacobinism. In a time of rapid political change, they acted as safe-conducts from one period into the next. The men, for example, whom contemporaries accused of complicity in the death of Lavoisier, might insure themselves after *thermidor* by publicizing accounts of their heroic service in the national war effort.[44] Legends also became important as the currency by which men forced to encounter each other for institutional reasons, might do so without overt hostility. The community of science could function as such only by avoiding giving utterance to the hostilities generated by the Terror.[45] Lastly, they enabled those bound together by enormous debts of gratitude to objectify and depersonalize their relationship. If open discussion was difficult, then the images of myth became doubly important, as providing an agreed language using well-known symbols such as the figure of Socrates, through which meaning could be conveyed.

V

The preservation of community was a problem which began to perturb contemporaries as soon as the Revolution began, and especially after 1792. They were well aware that science was more sensitive than any other form of intellectual activity to upheavals in its existence as a community.[46] For the Academy of Sciences in particular, changes in the constitution in 1785 and 1789 had already provoked heated debate between members.[47] But it was above all between 1789 and 1793 that the divided allegiances of the Academy between the Crown and the National Assembly resulted in serious conflict between factions and individuals.[48] These dissensions were brought to a head by Fourcroy's demand of 10 August 1792 that the Academy disown those members known to be *émigrés*. After 1792, flight and imprisonment lead to the physical dispersal of the Academy (Table II). More fundamentally, political and personal conflicts between members of the Academy became profound. The ultimate example of this is of course to be found in the circumstances surrounding the execution of the chemist Lavoisier, which have attracted more attention than any other single episode of the Terror as it affected the sciences. It is not my intention to reopen here the controversy surrounding the degree of complicity in Lavoisier's death of his pupils Fourcroy and Hassenfratz, and that of his colleagues in the Academy Berthollet and Guyton de Morveau.[49] I wish to emphasize here another point altogether. This is that the case of Lavoisier was a living reminder to the survivors of the Academy of the risks of patronage, one of the essential shapers of the scientific community. The issue of trust which always forms a reciprocal part of the relationship between patron and protégé, were blown up by such episodes under the Terror into dramas of life and death.

Against this sort of background, it becomes comprehensible that survivors of 1793–94 harped by 1795 on images of the reforging of scientific community. Letourneur, the member of the Directory who presided over the ceremonial opening of the Institut, reiterated this theme in his speeches.[50] A little later, one of the Permanent Secretaries of the Institut's First Class, the naturalist Cuvier, appealed to its members' tenderest sentiments by painting an affecting picture of the same first session. Men separated for months or years enquired after the experiences of those present, and mourned for those who could not be so.[51] Against the atomism of the Terror, he continued, the sciences oppose an ideal of true community:

> Fortunately, there exists in the midst of political associations an association of another kind, which tries to serve them all, but takes no part in their continual conflicts. The true friends of science, while as loyal to their country as the next man, are also united amongst themselves by the generous links which attach them to the great cause of humanity.[52]

The reality was less cosy. There was not only the dissatisfaction of the former members of the Academy of Sciences who had been excluded from the Institut to contend with,[53] there was also a new pressure from the state. Whilst condoling with the *savants* on their recent ordeal, Letourneur had left them in no doubt as to the direction of their future loyalties: "Citizens, if there are

103

still the wicked to punish, there are also the incredulous to convince, errors to combat, hatreds to soothe. . . . The law, which protects you, in return counts on your influence."[54] Liberation from the Jacobin Terror was not the equivalent of the liberation of culture from the political field. In this new era of mass politics, state power and individual isolation, the first era of 'big government' in French history, the manipulation of culture and the producers of culture of all kinds, was important as never before.[55] The new pressure from the state was different from that imposed by the monarchy before 1789, in virtue of the status of the Academy of Sciences as a corporation by royal charter and under the direct patronage of the Crown, and often employed by it to report on technical questions within its competence. Napoleon's was the first regime consciously to try to integrate the producers of culture into its new ruling élite. Earlier, with the abolition of the monarchy, and the Jacobin formation of the ideal of the nation, institutions themselves changed. They were no longer royal, but national, and as such bound up with the whole political field involved in the shifting ideal of the nation.

The individual members of the Institut were also more personally exposed to the pressures which Napoleon might choose to place upon their institution. There were many such as Cels, Le Monnier and Cousin, who, having been financially independent before 1789, had lost positions and income in the Revolution, and became dependent on the practice of science and their salaries as members of the Institut in a way they had never been before.[56] To later historians of science this might seem like a pre-condition for 'professionalization'; at the time it must have felt more like an increased vulnerability to the institutionalized pressure of the state.

It should not by now be necessary to spell out our agreement with the argument of Roger Hahn that the period of the Terror marked a turning point for the sciences, and not merely institutionally but also psychologically.[57] But we also point out that the period of the Terror, particularly for future members of the Institut who had not been members of the Academy of Sciences, could also strengthen vocational commitments rather than weakening them. For men like Bosc, Lacepède and Ramond, imprisonment and the enforced retreat into nature were crucial experiences in renewing, rather than undermining, their sense of vocation.

The closure of the Academy of Sciences, seen as a crucial event in this evolution of self-consciousness by many historians, in fact was of comparatively little importance in weakening some scientific vocations, just as it was in strengthening them in others. The astronomer Bailly suffered an especially ironic fate. At the outset of the Revolution, he was guided by the conviction that this great movement was destined to be led by the *savants*.[58] However, the very pace of events during his tenure of the office of mayor of Paris, culminating in the massacre of the Champ de Mars on 14 July 1791, ended in the obliteration of his vocational identity. He confessed to his former colleague in the Academy, the astronomer Lalande, ". . . that a torrent had passed over his head and had carried away all his ideas of science; he could hardly remember that he had ever been an astronomer".[59] Another well-documented case of vocational collapse is that of Cassini, hereditary Director of the Paris Observatory. The closure of the Academy undoubtedly affected him deeply. But in 1793 he also experienced the revolt of his pupils at the Observatory, his exclusion as a Royalist from the Directorship, and finally arrest and imprisonment. Meanwhile, the Jacobin government remodelled the Observatory's constitution so as to take away supreme power from the Director and distribute it 'democratically' among the rest of its employees.[60] For Cassini, this was a crisis of tremendous proportions. It shattered the connection of his family with the Observatory, a connection which had lasted for four generations. It expelled him from his home. It exposed the ingratitude of his pupils. The collapse of his authority completed the work begun by the attack on his old-fashioned observational astronomy by the new celestial mechanics of Laplace. In prison, Cassini wrote verses which summed up his state of mind:

> Dans ce triste séjour où gémit l'innocence,
> Que m'ont servi mon nom, les talents, la science?
> A prolonger le temps de mon captivité.
> En armant contre moi l'envie et l'ignorance.
> Hélas il est donc vrai: tout n'est que vanité.[61]

His entire vocation disappeared and Cassini abandoned astronomy altogether on his release from prison. His election to the First Class of the new Institut in December 1795 caused him great unease, and he resigned from it on 18 January 1796. He declined a nomination to the Bureau des Longitudes, the body administering the formerly independent Observatory. The most he would accept was non-resident membership of the Institut.[62] Most of the rest of his life was spent in local administration, in retreat into rural *otium*, and in the writing of works on his family history, and on the martyrology of the Terror. In Cassini's case, his vocation, authority, residence, family prestige, and research programme, as well as patronage power, all collapsed at the same time, and did so not only because of the closure of the Academy but far more because of the generalized attack on the old élites that the Terror licensed, provided the machinery for, and almost turned into a patriotic duty.

Hahn also argues that the closure of the Academy was a crucial stage in the relation between science and the state. Using a modified version of the 'mobilization of the savants' thesis discussed above, Hahn argues:

> As a result of the institutional changes that had taken place in August and September of 1793, the academician had been fully transformed into a public functionary. . . . He was led to abandon his primary allegiance to fellow scientists and to the scientific enterprise by his total absorption in governmental affairs.[63]

By using terms such as 'the academician' this passage leaves us with the impression that most members of the Academy had become involved with scientific work for the government. But as we have seen, such work was in fact the preserve of a relatively small number of men. How can this alleged major shift in values apply to the far greater numbers who experienced exile, ruin, and imprisonment?

Nor were those who were employed by the government the subordinates pictured by Hahn. Fourcroy and Guyton were to belong to the Committee of Public Safety itself, after *thermidor*.[64] Monge had been Minister of the Navy in 1792; Hassenfratz had precipitated the fall of the Gironde.[65] These men did not merely respond to changed circumstances, they actively sought to change them for themselves. They precipitated themselves into the political world to strengthen their own positions at a time when no man's future was secure, and, as all around the bonds of patronage were loosened, to strengthen theirs.

VI

We may now have demonstrated our substantial agreement with Hahn's major thesis on the importance of the Terror for the community of science. Much more can be said, however, in conclusion, about this period. So many men entered the first class of the Institut in 1795 without having been members of the Academy of Sciences that the shock of August 1793 cannot be invoked to explain the whole of their reactions. The Academy had certainly been of the highest importance, as Lavoisier pointed out, in providing an arena within which the substantive advance of science could be validated. At the same time, however, the scientist had always had another set of roles to fall back on, roles which the Academy certainly endorsed, but which could not be acted out within it: that of the *savant* as the field worker, engaging directly with nature far from the distractions of the corrupting human social and political world. This was an ideal which was in many cases strengthened by the circumstances of the Terror, rather than limited by some experience of bureaucratic involvement. In the rare cases where bureaucratic involvement took place, I believe that we are not dealing with the mobilized *savants* beloved of Aulard and Mathiez, and partially adopted by Hahn, but with a group of men and their protégés, relatively small in number, who entered political life as a means of self-advancement, and who did not visibly experience the disjunction between their political and scientific roles ascribed to them by Hahn. Where such disjunctions are found it is among the victims of the Terror, such as Ramond and Lacepède, in the men who survived imprisonment and upheaval by ever after keeping the stabilizing contemplation of nature in a separate compartment from their public involvements.

The consequences of the Terror for the sciences were many and complex, and can only be lightly sketched here. They went far beyond the disruption of research programmes and patronage relationships. This was a period of acute crisis between the former élites of the old regime and the new Jacobin state. This crisis is the underlying concern of the majority of the legends generated at this time in relation to the sciences and to individual scientists.[66] Science also became more vulnerable to attacks from outside the élite. It is a commonplace in contemporary statements that it was 'the mob', inspired by the Jacobins, who had been the worst enemies of France's intellectual élite.[67] The drive to create an élitist, highly technical and epistemologically independent science which marks so much of the nineteenth century in France, can be traced back to a fear of pressure and judgment both from the new, intrusive 'big state', and from the non-expert 'mob'. Both these pressures also increased the temptation to the individual scientist to find security in the endorsement of the strongest, Stoic, forms of retreat into the self, as an essential part of scientific vocational ideology.[68] Although it is probably too much to say that the Terror provided one of the pre-conditions for the professionalization of science in France, yet the men who went through it certainly came out with a different and closer relationship to their sense of vocation. The fates of Cassini and Bailly had shown the disintegration which awaited those who could not make vocation the centre of their being. Against the pressure from the state and from the semi-educated of which the Terror was only the highest power, personal commitments to science hardened. The Terror turned interest into vocation, and vocation into life-saving and life-committing passion.

Table I Members of the Academy of Sciences dying from any cause between 4 September 1792 and November 1795; 10 deaths directly related to the Terror[69]

Amelot de Chaillou, Antoine-Jean, 1732–20 April 1795. Died in the Luxembourg prison, Paris.

Bailly, Jean-Sylvain, 1736–12 November 1793. Guillotined on the Champ de Mars.

Bertin, Henry-Léonard, 1720–16 September, 1792. Died at Spa.

Bochart de Saron, Jean-Baptiste, 1730–20 April 1794. Guillotined at Paris.

Condorcet, Jean-Antoine Nicolas, 1743–29 March 1794. Committed suicide at Bourg-la-Reine near Paris to avoid trial and execution.

Cornette, Claude Melchior, 1744–11 May 1794. Died in Rome.

Demours, Pierre, 1702–23 May 1795.

Dietrich, Phillippe Frédéric, 1748–28 December 1793. Guillotined in Paris.

Dionis du Séjour, Achille-Pierre, 1734–22 August 1794. Died at Angerville.

La Rochefoucauld, Louis-Alexandre, 1743–4 September 1792. Assassinated at Gisors.

Lavoisier, Antoine-Laurent, 1743–8 May 1794. Guillotined in Paris.

Le Gentil de la Galaisière, Guillaume-Joseph, 1725–22 October 1792. Died in Paris.

Loménie de Brienne, Etienne Charles de, 1727–19 February 1794. Died at Sens.

Machault d'Arnouville, Jean-Baptiste de, 1701–12 July 1794. Died in the prison of the Madelonettes, Paris.

Malesherbes, Chrétienne Guillaume de Lamoignon, 1721–22 April 1794. Guillotined at Paris.

Mesnard de Chousy, Didier François René, 1729–18 April 1794. Guillotined in Paris.

Meusnier de la Place, Jean Baptiste, 1754–13 June 1793. Killed during the siege of Mayence.

Perronet, Jean-Rodolphe, 1708–27 February 1794. Died in Paris.

Petit, Antoine, 1722–21 October 1794. Died at Orléans.

Vicq d'Azyr, Félix, 1748–20 June 1794. Died in Paris.

Note: Of these men, Amelot, Bertin, Bochart, La Rochefoucauld, Loménie de Brienne, Machault, Malesherbes and Mesnard were Academiciens, the rest 'working members'.

Table II Members of the Academy of Sciences and of the Institut out of Paris from 1793[70]

Angivillier: emigrated 1792, died in Hamburg 1809 without returning to France.

Bailly: in refuge at Niort, July–September 1792; then imprisoned in Paris until his execution.

Baumé: left Paris 1793–95.

Beautemps-Beaupré: 1791–95, member of Entrecasteaux's expedition in search of la Pérouse; see also Rossel.

Borda: 1793–94, shared Coulomb's retreat at Blois.

Bosc: fled Paris after the fall of the Gironde in June 1793. In America 1795–1800.

Bougainville: left Paris for Coutances, 1792. Imprisoned August 1793 until *thermidor*.

Broussonnet: imprisoned in Montpellier October 1793, escaped to Spain, 1794, returned clandestinely to Paris 1795, consular posts abroad 1797–1800.

Cassini: imprisoned in Paris February–September 1794. Thereafter at Thury in Normandy.

Chabert: emigrated.

Claret de Fleurieu: arrested 1792, released after *thermidor*.

Coulomb: left Paris July 1793, returned 1795. See Borda.

Cornette: emigrated.

Cousin: imprisoned in Paris, January 1793, for eight months.

Delambre: frequently absent from Paris on work connected with the measurement of the meridian.

Desmarets: imprisoned for an unspecified amount of time during 1793.

Dietrich: in Strasbourg as Mayor, 1790, imprisoned in Paris 1792–94; guillotined.

Dolomieu: at La Roche-Guyon, 1792–94.

Huzard: part of army veterinary service: frequently with armies.

Lacepède: at Leuville, 1793–94.

Laplace: at Melun, 1793–94.

Lassus: emigrated to Rome, 1790–94.

Latreille: imprisoned in Paris, November 1793–January 1795.

Lévêque: at Nantes (naval school of hydrography).

Méchain: in Spain, 1792–95, to complete work on measurement of meridian.

Olivier: left Paris November 1792, returned 1798.
Palisot: in Africa and the West Indies, 1786–98.
Parmentier: on 'official mission' in the Midi, 1793–94.
Pelletan: army medical service.
Percy: army medical service.
Ramond: left Paris, 1792, imprisoned near Tarbès three times 1792–94.
Rochon: in Brittany, 1792–95.
Rossel: on Entrecasteaux's expedition 1789–95, prisoner in England, 1795–1811.
Sabatier: with the Armée du Nord.
Sage: imprisoned for several months in 1793.
Sané: at Brest on naval service, 1793–95.
Tenon: retreated to country 1793–95.
Tessier: in Normandy, 1793–95.

Table III

*(i) Full members of the Academy of Sciences who became
    members of the First Class of the Institut:*[71]

Adanson, Berthollet, Borda, Bossut, Brisson, Cassini, Coulomb, Cousin, Darcet, Daubenton, Delambre, Desfontaines, Desmarest, Duhamel, Fourcroy, Haüy, Jeurat, Jussieu, Lagrange, Lalande, Lamarck, Laplace, Legendre, Le Monnier (Pierre), Le Roy, L'Héritier de Brutelle, Méchain, Messier, Monge, Pelletier, Périer, Portal, Rochon, Sabatier, Sage, Tenon, Tessier, Thouin, Vandermonde.

*(ii) Those elected to the Institut who had not belonged to the
    Academy of Sciences:*[72]

Arago, Ampère, Beautemps, Berthoud, Biot, Bosc, Bouvard, Brongniart, Broussonet, Burckhardt, Carnot, Cels, Chaptal, Charles, Claret de Fleurieu, Corvisart, Cuvier, Deschamps, Desessarts, Deyeux, Gay-Lussac, Geoffroy St Hilaire, Gilbert, Girard, Hallé, Huzard, Lacepède, Lalande (Michel), Lassus, Latreille, Lefèvre-Ginau, Lelièvre, Malus, Mirbel, Molard, Olivier, Parmentier, Pelletan, Percy, Pinel, Poinsot, Poisson, Prony, Ramond, Richard, Rossel, Sané, Sylvestre, Thenard, Vauquelin, Ventenat, Yvart.

Table IV   The debts of the Terror[73]

Bailly: saved classicist Dussaulx from mob.
Borda: spoke for Lavoisier.
Bory: secured release of Latreille.
Bosc: hid Girondin Roland, became legal guardian of his daughter, hid future
    Director Larévellière-Lépaux.
Brongniart, Alexandre: aided Broussonet's escape to Spain.
Carnot: warned Prony of imminent arrest.

Cels: saved l'Héritier from arrest.

Cuvier: sheltered Tessier.

Daubenton: attempted to save Haüy from imprisonment.

Desfontaines: saved l'Héritier from execution; visited Latreille in prison; spoke for Haüy.

Dolomieu: protested against the murder of La Rochefoucauld.

Fourcroy: saved Darcet and Chaptal from imprisonment, intervened on behalf of Sage, Desault and Brongniart.

Geoffroy St Hilaire: saved Haüy from imprisonment, spoke for Daubenton, hid the poet Roucher.

Gilbert: supported those persecuted because of their links with Malesherbes.

Guyton: helped to save Chaptal from arrest due to his links with the revolt in Lyons.

Hallé: attempted to save Lavoisier and Malesherbes.

Haüy: spoke for Lavoisier.

Lalande, Jérôme: saved the Abbé Garnier and Dupont de Nemours from arrest.

Leroy: spoke for Lavoisier.

Pinel: sheltered Condorcet in Bicêtre.

Sage: spoke for Lavoisier.

Thouin: visited Ramond in prison; spoke for Lavoisier and Haüy, sheltered La Révellière in the Muséum.

Notes

1 The most convincing recent interpretation of the significance of the Terror is François Furet, *Penser la Révolution française* (Paris, 1978), translated as *Interpreting the French Revolution* (London, 1981).

2 François-René de Chateaubriand, *Mémoires d'outre-tombe* (Paris, 1958), 165. Unless otherwise indicated, all translations in the following pages are the author's. The ideology justifying the use of terror to create a republic of virtue is examined in Pierre Trahard, *La sensibilité révolutionnaire* (Paris, 1936).

3 Based on data drawn from Institut de France, *Index biographique de l'Académie des sciences du 22 décembre 1666 au 1 octobre 1978* (Paris, 1979). The period of the Terror is not treated in James E. McClellan III, "The Académie Royale des Sciences, 1699–1793: A statistical portrait", *Isis*, lxxii (1981), 541–67.

4 C. C. Gillispie, "The *Encyclopédie* and the Jacobin philosophy of science: A study of ideas and consequences", in M. Clagett (ed.), *Critical problems in the history of science* (Madison, Wis., 1959), 255–89; L. P. Williams, "The politics of science in the French Revolution", *ibid.*, 291–308; Roger Hahn, "Elite scientifique et démocratie politique dans la France révolutionnaire", *Dix-huitième siècle,* i (1969), 229–35.

5 Roger Hahn, *The anatomy of a scientific institution: The Paris Academy of Sciences, 1666–1803* (Berkeley, Los Angeles and London, 1971), 252–73.

6 M. J. Sydenham, *The first French republic: 1792–1804* (London, 1974), 3–25; R. R. Palmer, *Twelve who ruled* (Princeton, 1941); M. Mortimer-Ternaux, *Histoire de la Terreur, 1792–1794, d'après des documents authentiques et inédits* (8 vols, Paris, 1862); Marc Bouloiseau, *La république jacobine, 10 août 1792–9 thermidor II* (Paris, 1972), 106–23, 171–212.

7 Daubenton, the naturalist, dared not use his carriage, lest local sans-culottes interpreted this as an anti-democratic gesture; see the letter of Daubenton to Lavoisier quoted in F. Kafker, "Les encyclopédistes et la terreur", *Revue d'histoire moderne et contemporaine,* xiv (1967), 284–95, p. 289.

8 Donald Greer, *The incidence of the Terror during the French Revolution: A statistical interpretation* (Harvard, 1935), 26–27.

9 Georges Kersaint, "Antoine-François Fourcroy, 1755–1809: sa vie et son oeuvre", *Mémoires du Muséum national d'histoire naturelle,* série D (sciences physico-chimiques), ii (1966), 1–296, p. 31; Joseph-Jérôme Le François de Lalande, "Eloge de Bailly", *Décade philosophique,* iv, (1795), 321–30, p. 329; Benjamin Constant, *Des effets de la Terreur* (Paris, an V – 1797), 35; Alfred Lacroix, *Déodat Dolomieu: sa vie avantureuse: sa captivité: ses oeuvres: sa correspondence* (2 vols, Paris, 1921), i, p. xxvii. For the moderns, Mortimer-Ternaux, *op. cit.* (ref. 6), i, 8, gives 20 June 1792 as the opening date of the Terror; Kafker, *op. cit.* (ref. 7), 285, gives 31 May; Sydenham, *op. cit.* (ref. 6), 312, gives 5 September 1793.

10 Max Frey, *Les transformations du vocabulaire français à l'époque de la révolution, 1789–1800* (Paris, 1925), 187.

11 *Index biographique* (ref. 3). Such information proved difficult to locate for the following: Bory, Bossut (i.e. two out of the forty-eight working members of the Académie des sciences), La Billardière. Only resident members have been considered, and in all tables Bonaparte has been excluded from consideration as a member of the First Class of the Institut. With these exceptions, all the members of 1793 appear in the tables.

12 Delambre laid the blame for these purges at the door of Prieur de la Côte d'Or, Carnot's assistant in the Committee of Public Safety: Kersaint, *op. cit.* (ref. 9), 64–71.

13 L. F. A. Maury, *L'ancienne Académie des sciences* (Paris, 1864), 333. Comparison with the original members of the Second Class of the Napoleonic Institut shows that twenty of the forty-two resident members and thirteen of the forty-two non-resident members had their careers severely disrupted due to the Terror. Out of the total of eighty-four, five emigrated, fourteen were imprisoned, eight proscribed and denounced, and six spent this period in hiding. Such men were thus just as vulnerable as were those engaged in the physical and life sciences. Martin Staum, "The class of moral and political sciences, 1795–1803", *French historical studies,* xi (1980), 371–97. David L. Dowd, "The French Revolution and the painters", *French historical studies,* i (1959), 127–48, shows that the level of disruption was far less in the case of former members of the Académie royale de peinture et de sculpture. I would like to thank Professor Staum at this point for his valuable comments on earlier versions of this paper.

14 Maurice P. Crosland, "The French Academy of Sciences in the nineteenth century", *Minerva,* xvi (1978), 73–102, p. 76: "The early nineteenth century membership of the Institut was continuous with the pre-revolutionary membership of the Academy"; René Taton, "The French Revolution and the progress of science", *Centaurus,* iii (1953), 73–89, p. 80: "The majority of scientists . . . came through the revolutionary period without any great difficulty"; Joseph Fayet, *La révolution française et la science, 1789–1795* (Paris, 1960), 432.

15 Quoted in Lacroix (ed.), *op. cit.* (ref. 9), p. xxvii: ". . . une telle impression de terreur, de crainte, que je n'en suis encore délivré. L'inquiétude, la méfiance y sont empreintes sur toutes les figures, on redoute de se parler, on n'ose hasarder une opinion, l'egoisme exerce un empire absolu sur tout le monde; il n'y a plus de liaisons, ni amitié, tout est faction, tout est parti."

16 Kersaint, *op. cit.* (ref. 9), 32, quoting from Fourcroy's autobiography: "J'ai manqué de pain pendant dix mois. J'ai vécu, moi et cinq personnes chez moi, de pommes de terre . . . je ne voyais presque personne encore; je refusais les invitations, les dîners, j'étais comme à mon ordinaire étranger à tout parti."

17 Charles B. Paul, *Science and immortality: The éloges of the Paris Academy of Sciences, 1699–1791* (Berkeley, Los Angeles and London, 1980), 92–109; D. Outram, "The language of natural power: The *éloges* of Georges Cuvier and the public language of nineteenth-century science", *History of science,* xvi (1978), 153–78.

18 Henri Dehérain, "Une autobiographie du Baron Ramond", *Journal des savants* (1905), 121–9: "Dans cette longue solitude, dans ce profond silence, toujours en présence de l'échafaud qui m'attendait, que pensez-vous que j'aye fait? Me souvenir? Non: oublier. Aussi suis-je sorti de prison sans rancune, dépouilli de tout, vêtu de vieux haillons, marchant sans souliers et vivant de quelques sous qu'on avait oublié de me voler, je m'en suis pris aux évenements, non aux hommes, j'ai vaincu le sort par la patience, cherché mes consolations dans l'étude de la nature, et regardant toujours en avant, jamais en arrière, j'ai reconquis ma position sociale à force de travail et d'économie. . . . Ah! le beau livre que je vous ferais, nouveau Robinson que j'ai été, sur les miracles de patience, du travail et de l'économie, sur le noble orgueil de l'honnête homme qui n'a de salut qu'en soi, et se suffit à lui-même."

19 Dehérain, *op. cit.* (ref. 18), 128.

20 Claude Perroud, *Mémoires de Mme Roland: nouvelle édition critique: contenant des fragments inédits et les lettres de prison* (2 vols, Paris, 1905), ii, 455–6: "je m'habillai à la sans-culotte, travaillai à la terre, au bois . . . fis moi-même ma cuisine. . . . J'employais les autres jours à l'histoire naturelle, aux travaux manuels, et à la chasse. Ne pouvant suppléer aux livres qui me manquaient, j'entrepris une description, accompagné de figures, de toutes les araignées de la forêt, comme la partie la moins connue." This work, dated September 1793, is MS 872 of the Central Library of the Muséum national d'histoire naturelle, Paris, and is entitled "Araignées de la forêt de Montmorency décrites et déssinées pendant que j'étais caché à Radegonde".

21 Roger Hahn, "L'autobiographie de Lacepède retrouvée", *Dix-huitième siècle,* vii (1975), 49–85, p. 69: "Assis sur les ruines que environment la haute tour de Montlhéry, dominant sur un pays immense . . . ou couché sur un gazon fleuri, à l'ombre des peupliers inspirateurs et sur les bords du grand étang de Marcoussis, ou me promenant sous les voûtes de verdure formées par les vastes et solitaires forêts qui courronnaient les montagnes autour de cet étang, j'aimais à méditer sur les admirables effets de la puissance de la nature, sur la sublimité de ses lois . . . j'oubliais le monde, je ne voyais plus que l'univers."

22 Hahn, *op. cit.* (ref. 21), 78: ". . . une peine plus forte qu'on ne pourrait le croire, et à laquelle je n'ai jamais pu m'accoutumer."

23 Quoted in Trahard, *op. cit.* (ref. 2), 134–5: "Dans les débats orageux, on s'indignait de ton absence et ton silence: toi, tu parlais de la campagne, des délices de la solitude et de la paresse. . . ."

24 Daniel Roche, *Le siècle des lumières en province: académies et académiciens provinciaux, 1680–1789* (2 vols, Paris, 1978), i, 391.

25 Painting listed in Jean Starobinski, *1789: les emblèmes de la raison* (Paris, 1973), 81, 110; specifically alluded to here are David's "Socrates taking hemlock", and Canova's "Criton closes the eyes of Socrates". For an extension of this analysis, see Norman Bryson, *Word and image: French painting of the Ancien Régime* (London, 1981), 233–4.

26 Lalande, *op. cit.* (ref. 9), 329: "Il préfera l'example de Socrate"; Millin de Grand-maison, "Elémens d'histoire naturelle", *Magasin encyclopédique,* iii (1795), 14–22, pp. 21–22.

27 Jean-Dominique Cassini, *Mémoires pour servir à l'histoire des sciences, et à celle de l'observatoire royale de Paris . . . et des éloges de plusieurs académiciens morts pendant la révolution* (Paris, 1810), 388; Millin, *op. cit.* (ref. 26); Jean-Baptiste Biot, *Essai sur l'histoire générale des sciences pendant la révolution française* (Paris, 1803), 54; Emmett Kennedy, "Destutt de Tracy and the unity of the sciences", *Studies in Voltaire and the eighteenth century,* clxxi (1977), 223–39.

28 See also Georges Pouchet, *Les sciences pendant la Terreur d'après les documents du temps* (Paris, 1896). Furet, *op. cit.* (ref. 1), discusses the historiography *passim,* as does Roger Hahn, "The problems of the French scientific community, 1793–95", *Actes du douzième congrès internationale de l'histoire des sciences* (1974), iii B, 37–40.

29 Albert Soboul (ed.), *Oeuvres d'Henri Grégoire* (14 vols, Paris, 1977), ii, 258–391.

30 *Ibid.*

31 Eugène Despois, *Le vandalisme révolutionnaire: fondations littéraires, scientifiques et artistiques de la Convention* (Paris, 1868).

32 Honoré Riouffe, *Mémoires sur les prisons* (2 vols, Paris, 1823), i, ll: "Les comités révolutionnaires, au lieu de diriger leur feu vers un certain but, faisaient un feu qui écartait. Des petites villes entières se traînaient à l'échafaud; mais c'est le marchand qui dénonciait le marchand, et tous deux étaient arrêtés par celui qui avait été leur ouvrier. C'était des haines de voisin, des jalousies de profession, qui prenaient tout leur essor sous un masque révolutionnaire."

33 Grégoire's efforts were all the more necessary as there were many who were unwilling to absolve individual scientists from responsibility for the events of the Jacobin Terror. Lacepède's bland obituary of Vandermonde was for example criticized by the *Décade philosophique,* ix (1796), 152: ". . . ni les talents, ni la science ne pourront soustraire au blâme et à la honte, la mémoire de tous ceux qui ont été, de quelque manière que ce soit, les échos ou les complices de cette horrible tyrannie."

34 Biot, *op. cit.* (ref. 27), 1: "Le despotisme révolutionnaire leur donna une existence politique, il s'en servit pour inspirer de la confiance au peuple, pour préparer des victoires et gagner des batailles."

35 Albert Mathiez, "La mobilisation des savants en l'an II", *Revue de Paris,* xxiv (1917), 524–65; Taton, *op. cit.* (ref. 14); Hahn, *op. cit.* (ref. 5), 292.

36 Constant, *op. cit.* (ref. 9), 22–28.

37 *Ibid.,* 30–31: "C'est la frénésie de 1794 qui fait abjurer par les hommes foibles ou aigris, les lumières de 1789."

38 Georges Bouchard, *Guyton-Morveau: chimiste et conventionnel* (Paris, 1938), 333–5; Fayet, *op. cit.* (ref. 14), 252.

39 C. Richard, *Le comité de salut publique et les fabrications de guerre sous la Terreur* (Paris, 1921), 56–84, 208–12, 663–84; Kersaint, *op. cit.* (ref. 9), 259; Arthur Birembaut, "Précisions sur la biographie du mathématicien Vandermonde, et de sa famille", *72e Congrès de l'Association française pour l'avancement des sciences,* lxxii (1953), 530–3.

40 Louis Audiat, "Le mot de Bailly allant à l'échafaud", *Revue des questions historiques,* xx (1876), 544–53.

41 James Guillaume, "Le berger Daubenton: encore une légende contre-révolutionnaire", *La révolution française,* xlii (1902), 385–98. Guillaume contradicts the account in Georges

Cuvier, *Receuil des éloges historiques* (3 vols, Paris and Strasbourg, 1819–27), i, 27–80, which is, however, nearer to contemporary views; Lacroix, *op. cit.* (ref. 9), 50–51, cites a letter from the geologist Faujas to this effect.

42 James Guillaume, "Un mot légendaire: la république n'a pas besoin des savants", *La révolution française,* xxxviii (1900), 385–99.

43 Albert Mathiez, *Robespierre terroriste* (Paris, 1926).

44 Georges Laurent, "Un mémoire historique du chimiste Hassenfratz", *Annales historiques de la révolution française,* i (1924), 163–4, shows how Hassenfratz emphasized his war work, and was strangely silent on his political career, which included the speeches which precipitated the fall of the Gironde.

45 It was on these grounds that the chemist Armand Seguin advised Lavoisier's widow not to accuse publicly those she considered responsible for his death: Edouard Grimaux, *Lavoisier* (Paris, 1899), 332.

46 Lavoisier commented, "Science is not like literature. The man of letters finds in society all the elements he needs to develop his talents . . . he depends upon no-one. The same is not the case in the sciences. Most of them cannot be pursued with success by isolated individuals." Quoted in Hahn, *op. cit.* (ref. 5), 234.

47 Roger Hahn, "L'académie royale des sciences et la réforme de ses statuts en 1789", *Revue d'histoire des sciences,* xviii (1965), 15–28.

48 Such conflicts were not unique to the Académie des sciences. Morellet, *Mémoires* (2 vols, Paris, 1822), i, 424, relates that conversation inside the Académie française "était dégénérée en querelle habituelle".

49 Grimaux, *op. cit.* (ref. 45), 266–376; Bouchard, *op. cit.* (ref. 38), 335–7; Roger Hahn, "Fourcroy, advocate of Lavoisier?", *Archives internationales de l'histoire des sciences,* xii (1959), 285–8.

50 Ernest Maindron, *L'Académie des sciences: histoire de l'Académie: fondation de l'Institut national: Bonaparte membre de l'Institut national* (Paris, 1888), 174.

51 Cuvier, *op. cit.* (ref. 41), i, 305: ". . . ce fut une impression ineffaçable . . . de ces larmes de joie, de ces questions réciproques, et empressés sur leurs malheurs, leurs retraites, leurs occupations, de ces douloureux souvenirs de tant de confrères victimes des bourreux. . . ."

52 *Ibid.,* 334: "Heureusement il existe au milieu des associations politiques une association d'un autre ordre, qui cherche à les servir toutes, mais qui ne prend point de part à leurs continuelles dissensions. Les véritables amis des sciences aussi dévoué à leur patrie qu'aucune autre classe d'hommes, sont encore unis entre-eux de ces mêmes liens généraux qui les rattache à la grande cause de l'humanité."

53 Of the original members of the First Class, seventeen out of twenty had belonged to the Academy of Sciences. Of the twenty whom they in turn elected, only nine had belonged to the Academy. See Table III.

54 Maindron, *op. cit.* (ref. 50), 174: "Citoyens, s'il est encore des méchants à punir, il est aussi des incredules à convaincre, des erreurs à combattre, des haines à désarmer. . . . La loi, qui vous protège, compte à son tour sur votre influence."

55 This is the topic of the author's work (in preparation), *A dimension of power: The cultural élite of Napoleonic France.* For the lengths to which Napoleon was prepared to proceed to obtain conformity from the members of the cultural élite, see François Aulard, "Napoléon et l'athée Lalande", *Etudes et leçons sur la révolution française* (4 vols, Paris, 1904), iv, 303–16.

56 Perroud, *op. cit.* (ref. 20), 450–5; see entries for Bosc, Cels and Jacques Charles in L. G. Michaud, *Biographie universelle ancienne et moderne, nouvelle édition* (45 vols,

Paris, 1843–65). This point is discussed further in Dorinda Outram, "Politics and vocation: French science, 1793–1830", *The British Journal for the history of science,* xiii (1980), 27–43; the effects of role-changes during the Revolution are discussed in Marguerite Vergnaud, "Un savant pendant la révolution", *Cahiers internationaux de sociologie,* xvii–xviii (1945–55), 123–39.

57 Hahn, *op. cit.* (ref. 5), 252, 273.

58 Gene Brucker, *Jean-Sylvain Bailly, revolutionary Mayor of Paris* (Urbana, 1950), 79.

59 This anecdote is related twice in Jean-Baptiste Delambre, *Histoire de l'astronomie au dix-huitième siècle* (Paris, 1827), 563, 748. I have followed the latter version.

60 Charles-Joseph-Etienne Wolf, *Histoire de l'Observatoire de Paris de sa fondation à 1793* (Paris, 1902), 321–52.

61 Quoted in Jean-François Schlisteur Devic, *Histoire de la vie et des travaux scientifiques et littéraires de J. D. Cassini IV* (Clermont, 1851), 290–1.

62 Crosland, *op. cit.* (ref. 14), 76, uses Cassini, however, as an example of continuity between the Academy of Sciences and the Institut.

63 Hahn, *op. cit.* (ref. 5), 263–4.

64 James Guillaume, "Le personnel du comité de salut publique", *La révolution française,* xxxviii (1900), 297–309; Kersaint, *op. cit.* (ref. 9), 31, shows that the dates given by Fourcroy himself for his membership of the Committee are erroneous. It is worth remembering that the Committee of Public Safety was founded not by the Jacobins, but by the Girondins in April 1793, and remained in existence until November 1795, well after *thermidor*. The Revolutionary Tribunal itself was not abolished until May of that year.

65 Claude Perroud, *La proscription des Girondins* (Paris, 1917), 34.

66 Stresses in the public language of science become exceptionally strong in this period, and in that of the Directory and Empire, and betray the pressures which the newly defined public and political life of France placed upon the sciences. These issues are discussed in Outram, *op. cit.* (ref. 17).

67 Popular and élite science had of course frequently come into conflict before the Revolution: Robert Darnton, *Mesmerism and the end of the Enlightenment in France* (Princeton, 1968). For one characterization of the period, see Cuvier, *op. cit.* (ref. 41), i, 383: "Cette époque funeste où tout mérite personnel, toute prééminence indépendant était odieux à l'autorité, et où l'on ne permettait de louer que les oppresseurs de la patrie et leurs plus méprisables satellites."

68 Themes developed further in Dorinda Outram, *Career-making in post-revolutionary France: Georges Cuvier and the life of science, 1796–1832* (forthcoming, Manchester University Press, 1984), ch. vi.

69 Based on Institut de France, *op. cit.* (ref. 3). From 1785 the total number of working Academicians was forty-eight, distributed in eight classes of six members each.

70 Based on *ibid.*; Michaud, *op. cit.* (ref. 56); L. Bobé, *Mémoires de Charles Claude Flahaut, comte de la Billarderie d'Angivillier* (Copenhagen and Paris, 1933); Roger Hahn, "Quelques nouveaux documents sur J. S. Bailly", *Revue d'histoire des sciences,* viii (1955), 338–53; Jean Mascart, *La vie et les travaux du Chevalier Jean Charles de Borda* (Lyons, 1919); Perroud, *op. cit.* (ref. 20); Devic, *op. cit.* (ref. 61); Charles Stewart Gillmor, *Coulomb and the evolution of physics and engineering in eighteenth century France* (Princeton, 1971), 75; Albert Mathiez, "Un complice de Lafayette, Friedrich Dietrich: d'après les documents inédits", *Annales révolutionnaires,* xii (1920), 389–408; Lacroix, *op. cit.* (ref. 9); Hahn, *op. cit.* (ref. 21); Paul Dupuis, "Pierre-André Latreille",

Annual review of entomology, xix (1974), 1–14; Maurice Bouvet, "Parmentier fut-il inquiété sous la révolution?", *Revue d'histoire de pharmacie,* xlviii (1960), 419–23; Dehérain, *op. cit.* (ref. 18); Cuthbert Girdlestone, *Poésie, politique, Pyrenée: Ramond, 1755–1827, sa vie, son oeuvre littéraire* (Paris, 1968), 225–39; Cuvier, *op. cit.* (ref. 41); M. Prévost and J. Roman d'Amat (eds), *Dictionnaire de biographie française* (12 vols, Paris, 1952–); Charles Coulston Gillispie (ed.), *Dictionary of scientific biography* (16 vols, New York, 1970–80). One hundred and two men were elected to the First Class of the Institut between 1795 and 1814. The Class had a membership of sixty at any one time.

71 Institut de France, *op. cit.* (ref. 3); a disputed membership is discussed in P. Dorveau, "Vauquelin fut-il membre de l'Académie royale des sciences?", *Revue d'histoire de la pharmacie,* no. 78 (1932), 57–60.

72 Institut de France, *op. cit.* (ref. 3). It is clear that the former Academicien Lemonnier, whose residence in Montreuil after 1793 disqualified him from full membership of the First Class, was still perceived as an integral part of its membership by the Institut; hence his éloge by Cuvier, an observance usually reserved for the resident members of the Class: Cuvier, *op. cit.* (ref. 41).

73 Brucker, *op. cit.* (ref. 58), 88–89; Perroud, *op. cit.* (ref. 20); Grimaux, *op. cit.* (ref. 45), 275; Fayet, *op. cit.* (ref. 14), 83; Cuvier, *op. cit.* (ref. 41), i, 257; Alfred Lacroix, "René-Just Haüy: sa vie et son oeuvre", *Bulletin de la société française de minéralogie,* lxvii (1944), 15–26; Lacroix, *op. cit.* (ref. 9), xxviii; Kersaint, *op. cit.* (ref. 9), 259; Théophile Cahn, *La vie et l'oeuvre de Etienne Geoffroy St Hilaire* (Paris, 1962), 18–22; Bouchard, *op. cit.* (ref. 38), 335–7; Constance de Salm, "Eloge historique de M. de Lalande", *Magasin encyclopédique,* ii (1810), 282–325, pp. 315–16; F. Kafker, *op. cit.* (ref. 7), xiv, 284–95; E. Gilbrin, "La ligne médicale des Pinel: leur aide aux prisonniers politiques sous la Terreur", *Histoire des sciences médicales,* xi (1977), 29–35; Pierre Grosclaude, *Malesherbes: témoin et interprète de son temps* (Paris, 1961), 690; Wolf, *op. cit.* (ref. 60), 352.

5

BEFORE OBJECTIVITY

Wives, patronage, and cultural reproduction in early nineteenth-century French science

From: Pnina Abir-Am and Dorinda Outram (eds.), *Uneasy Careers and Intimate Lives: Women in Science 1789–1972* (New Brunswick: Rutgers University Press, 1987), 19–30. © 1987 by Rutgers, the State University. Reprinted by permission of Rutgers University Press.

"They are never bored," wrote the composer A. E. M. Grétry of his female contemporaries, "because they are ceaselessly innovating."[1] He, like many others in early nineteenth-century France, was not short of ways to acknowledge and emphasize the debt cultural innovation in general owed women. Traditionally, history of science has been the least capable of all the histories of acknowledging the contribution of women either to its substantive or to its social development. In this essay I will assess that contribution for a period of history when there were so few women (just as there were relatively few men) gainfully engaged in science that women's contribution to science can only be examined through their interaction with the social organization of science, and with the value systems they might have been able to inject into it.

As I have argued elsewhere, this period is one in which science was as much organized through highly personalized patronage systems as it was through institutions.[2] Accordingly, it makes sense to examine women's role in that system, and in doing so extend our understanding of its operation from the traditional male dyad of patron and protégé to a new view of a triad including patron, patron's wife, and protégé.[3] Such a triad also performed far more complex functions than the simple transmission of power traditionally portrayed in accounts of the patron–client relationship. While most essays in this book implicitly or explicitly view marriage and career for women as reconcilable only with difficulty, this essay twists the perspective and explores marriage among scientific elites, and the female contribution to those marriage patterns, as actually constitutive of scientific ethos and the practice of scientific patronage.

Late eighteenth-century thinking on marriage and on the scientific vocation crisscrossed in many ways, some of which demonstrate the intimate dependence of scientific ideologies on the field of concepts surrounding them. Powerful currents of thought in this period rejected the world (*le monde*) and vaunted the simple, the natural, and the good as a means to the making of authentic, uncorrupted human beings.[4] A true sensibility came from such a rejection of the world and was also seen as an integral part of the image of the savant and the philosophe. Scientists themselves were continually enjoined to a direct contact with nature, and rejection of the corrupt world of intrigue and advantage, as the price of their ability to see nature as it was.[5] Both the clichés and the technical philosophy of the time thus tended to resolve the perceiver into that which was perceived. Scientific objectivity hardly made sense as a value against this ideology. The pure in heart were also the pure in eye. Against this background marriage fit in at best uneasily. Some could agree, like the naturalist L. Ramond, that marriage was the social institution that lay nearest to "nature."[6] For others, the inescapable companion of marriage, however neutralized by Rousseauist idealization, was *le monde:* the world of social connection and social leverage, and above all, as romantic ideas of marriage as companionship gained ground, of lifelong negotiation with another human being, with all of its implied threat to the blessed objectivity of the man of science. There were many who saw the marriage of the savant as nothing but that "perilous leap" against which Jean d'Alembert warned Joseph Lagrange.[7] And just as parental dismay traditionally attended the awakening of a scientific vocation, sure prelude to poverty, trial, and social marginalization, so parental alarm was easily triggered at the thought of marriage with a man of science. The well-known nervousness of the Marquis de Condorcet's future parents-in-law must have been paralleled in many other cases.[8]

Marriage, in short, with its aura of compromise and negotiation, with its inevitable links with all the world of social production, enjoyed an ambivalent relationship with the official ethos of science even when, as was true in this early period, there was little threat of professional competition from the side of the wife. While family was linked, as was the ethos of science itself, to ideals of retreat, purity, and sensibility, marriage in the real world disrupted that ideal person whom the man of science was supposed to be. In an age much given to the insistent construction of figures of representative virtue as guides to the business of living, this was a serious drawback, and one that, as we will see, was fully reflected in the marriages of individual men of science.

But precisely because of its ambiguous ideological setting, marriage could also perform a valuable mediating role. The wife typically took on roles essential to the real-world maintenance of scientific organization, such as the handling of protégés, thus preserving intact the claims of her husband to exert his "scientificness" entirely outside the tainted world of career making, patronage, and advantage.[9] At an even deeper level, it was one of the functions of the woman to reconcile yet another profound conflict in scientific life: the demands of the individual vocation and of its social implantation. As noted earlier, the discovery of a scientific vocation by a young man was hardly ever a cause of parental rejoicing in eighteenth-century France. Such conflicts were so frequent that they were almost an identifying mark of the true savant. Georges Cuvier said publicly: "It is with such similar conflicts over the choice of a career that the stories of many of our colleagues begin."[10] Such a rupture with the family forced the fledgling savant to confront an enormous paradox: the freedom to pursue innocent knowledge, and thus the freedom to become who that person authentically was, could occur only as the result of an act of archetypal guilt – the rejection of parental authority. The eighteenth century insisted strongly on submission to parental, and in particular paternal, authority as a moral duty analogous to submission to the will of the heavenly Father. Rejection of parental authority was thus in a true sense an act ushering in chaos, for it upset a chain of power that stretched from man to God. This was an ideology that the conservative state of Napoleon was no less concerned to preserve than had been the "divine right" monarchy of Louis XVI.

The only way out was to re-enter a family grouping, for only this could cancel the original act of betrayal of the young scholar's biological family.[11] The finding of a second family was always fused in contemporary accounts with adoption by an effective patron, the sine qua non of the effective implementation of an intellectual vocation, as it was, of course, of entry into any other form of public role. The young savant at this point is allowed to be the person he authentically is by the man who becomes at once his public spokesman and supporter, as well as his "good" father. Contemporaries agree on ascribing a transforming power to the moment of adoption by the patron.[12] Accounts telescope the grueling process of the search for an effective patron into a single miraculous moment of acceptance, a moment that wipes out former inadequacy and doubt at the same time it dissolves all the claims of the biological father over the young protégé. Eleuthère Du Pont de Nemours, for example, scarred by bitter quarrels with his father after the early death of his adored mother, credited François Quesnay with transforming him from an infant into a man; Louis Daubenton released the twenty-year-old Etienne Geoffroy St. Hilaire from parental authority by the sudden words "I have a father's authority over you."[13] The self-image of the protégé damaged by conflict with parents, and by the struggle for survival in a world that saw his vocation as of marginal importance to its great concerns, is reaffirmed in these encounters with an embodiment of intellectual authority in the precise lineaments of another human being.

Yet other contemporary accounts also strongly acknowledge the importance of fictive "mothers," whether single women or the wives of male patrons. These women are credited by contemporaries with a crucial role in the discovery and enforcement of vocation, which to my knowledge has never been examined in a more than anecdotal way by any historian of intellectual groups of this period. Here I can only begin to suggest what such a study should entail. But it does seem clear that we need an approach to the problem of the formation of patronage networks that is not exclusively tied to the two-person relationship of patron–protégé; we need to acknowledge the numbers of men who had no single obvious male patron, but who, like Georges Cabanis or André Morellet, remained linked with a female patron on a lifelong basis.[14] We need also to recognize that other male patrons were married to women to whom they often delegated some of the most vital parts of the operation of patronage, and that the protégés often established relationships with these women that were far more important to their development and advancement than were those with the husband; the circle around

Mme. de Prony, wife of the director of the Ecole Nationale des Ponts-et-Chausées (a prestigious college for civil engineers), is a case in point.[15]

In expanding our analysis in this way we also acknowledge contemporary ideas about the importance of women in the cultural field. Social and legal historians have recently emphasized the declining position of women in general under the Directory and Empire; in the early years of the Empire, legislation on marriage and the family was systematically abrogated in favor of a return to a patriarchal family system.[16] These arguments are important and valid, but they do not accord with other evidence from this period on the importance contemporaries attached to women's impact on the cultural field and on the shaping of individual character.[17] We should not mistake this emphasis for what it became in the later nineteenth century, when according women supreme importance in the realm of personal relations, home and family, and the consumption of cultural goods was strongly linked with efforts to restrict them to a purely privatized existence.[18] Culture was vitally important because it produced the symbols through which the public realm conducted the great debates of this period on the nature of power and the relationship between individual and state. Thus, to say that women had an important role in this field of culture is not to describe a trivialized group excluded from real participation in the public realm; it is to describe one of the substantive, innovative sectors of that realm at work.

Contemporaries were clear about the ways in which a relationship with a female patron, with its exposure to specifically female ways of exploring individual personality, was a powerful force in encouraging the young savant. The writer J. B. Suard, for example, stressed in his panegyric on his "fictive mother," Mme. d'Houdetot, how her particular way of getting to know people emphasized and displayed their strengths: "It seemed as if nature had given her an insight all her own; she found out quickly and surely the best that there was in any person she talked to, just as she did in any book that she read."[19] Or as Benjamin Franklin had earlier remarked to Mme. Helvétius: "In your company we are not only pleased with you but better pleased with one another and with ourselves."[20] This sentence neatly reveals how the enabling and reinforcing role of such women was not simply confined to individuals, but also affected the groups of their habitués, giving increased confidence and cohesion to this newest section of the French elite. Typically, groups of individuals linked to such women tended to form the social institution known as a salon. Unfortunately, in spite of the importance of the salon in promoting the cohesion of the intellectual elite, and in channeling individuals into the network of patronage relations, we possess no general account of the salon as a dynamic social institution for the period of the Revolution and the Empire. The fine works of Deborah Herz on Germany, Carolyn Lougée on seventeenth-century France, and Alan Charles Kors on the "d'Holbach coterie" have either concentrated on the history of one salon or have tried to write the salon into the history of women's intellectual emancipation.[21]

My aim is not to contradict these insights, but to suggest that salons need to be integrated into a full appreciation of their use as power bases both for the patron and his wife.[22] The salon was where the patron, or "father," first encountered his "sons," or protégés, through the medium of his wife, the fictive mother. At any one time it would contain individuals who were known and accepted members of the "family" and others, less familiar, who were on trial, waiting to be evaluated before their adoption was certain. We possess a detailed account of this process as it was carried out by Mme. de Prony, wife of engineer and physicist Gaspard de Prony. The account was written by her favorite fictive son, the economist and technologist Charles Dupin:

> Mme. de Prony, who could have chosen her friends from the highest ranks of society, yet showed no eagerness to receive the homage of men who had little to recommend them except birth and title. It was personal, intellectual merit, and, even more, merit of the heart, which she looked for in her friends above all else. . . . She had a sort of tact all too rare today, whereby she could assign to each person, from the first to the last of her intimates, the precise degree of esteem and consideration they deserved, without in the least seeming to deprive anyone of the courteous regard that was their due. This equality of treatment is the only way to make the world of society tolerable. Mme. de Prony could raise anyone to the precise rank in her esteem to which their age, merit, and social position entitled them, while never denigrating anyone. She loved above all to show her goodwill to young men who, with some hope of success, were just starting in their careers.[23]

The salon provides one of the strongest examples of how an informal social institution, dominated by women in their actual management, even if ultimately financed by the husband, and using the most subtle interpersonal discriminations as a guide to access to the public power of the patron himself, was in fact an integral part of the making of a career and the gaining of reputation. Henri Beyle, the contemporary expert on the interconnection between public and private politics, and himself a dedicated habitué of salons, made the point unequivocally in his autobiography:

> What a difference it would have made if M. Daru or Mme. Cambon had said to me in January 1800: Dear cousin, if you want to have an assured place in society, you must have twenty people who have a good reason to recommend you. So, choose a salon, go there every Tuesday (or whatever its day is) without fail, make absolutely sure that you are winning, or at the very least polite, to everyone who goes to that salon. You will be something in the world, you can even hope to win the favor of an amiable woman, when you are backed by two or three salons; and at the end of ten such years, these salons will open every door to you.[24]

The institution of the salon, crucial to the creation of the fictive family of protégés gathered around the patron and his wife, was also heavily dependent on the demography of the intellectual elite. The salon could only have functioned as it did within a given sort of family structure. There have, however, been no previous studies of patterns of family and marriage among the scientific elite.[25] To this end,

I have collected demographic information for a distinct sample defined by membership of the Institut de France between 1795 and 1815, in order to compare, with some crudeness, the resulting group of some individuals with information already available on the demographic behavior of other sections of the Napoleonic elite. Interesting differences emerge.

Possibly the largest single study of the Napoleonic elite is Louis Bergeron and Guy Chaussinand-Nogaret's exhaustive examination of the lists of provincial notables. From this emerges a picture of an elite characterized, as were the members of the Institut, by late marriage – around age forty-three for men, compared to twenty-eight in the general population.[26] In both the case of the notables and of the Institut members, differences in age between husband and wife tended to be large, often of more than ten years, as was to be expected if the group was to reproduce itself. Among the notables, many remained unmarried at age fifty. The Institut sample tends to reinforce these trends. The number of lifelong bachelors is even higher than among the notables, partly due to the many priests and former priests among the intellectual elite.[27] We need to modify this picture by recognizing the sizable numbers from the Institut group who chose to remain outside formal marriage, but produced children within more or less stable informal liaisons.[28] For many members of the Institut, having children within marriage was clearly low on their list of benefits to be gained from personal relations. This picture is reinforced when we consider the family size of those members who were legally married. The Institut scientists produced far fewer than the 2.83 children per marriage noted by Bergeron and Chaussinand-Nogaret among the notables, which is itself around half the average in the general population. The Institut sample is also remarkable for the high number of marriages within its ranks that produced no children at all.[29] If normal fertility in this group is assumed, deliberate choice against the production of children is clearly at work. All these factors mean that for the Institut members marriage was not the reproductive machine it appeared to be for many of the French elite in the later nineteenth century.[30]

Typical family structure in the Institut, on the other hand, facilitated, if not necessitated, the formation of the fictive family of protégés so strongly encouraged by the treatment, in the vocational ideology of the savant, of the patron as equivalent to the rejected and rejecting biological parent. The low numbers of children produced in the average Institut family meant that the fictive family had space to accumulate. Since the wife was not preoccupied with the care of many children, she had the time and emotional space to devote to her husband's protégés and, in doing so, to influence their careers.

The fictive family was also, it can be argued, a necessary social device for this specific elite, which is characterized by the wide disparities in income between its members and a lower average income than any other section of the elite.[31] Relatively few in this group seem to have married to remedy a deficiency of capital. In these circumstances, to have produced higher numbers of children would not only have meant higher investment in their early years, but also high single-time expenses, such as the provision of dowries for daughters. It was clearly far less expensive to maintain the fictive family. Without the fictive family, the demography of the Institut group would also be ineffective in ensuring transmission of cultural styles and research programs into the next generation.[32]

We have now entered an area in which the demographic evaluation of these marriages has to be reinforced by a qualitative study. It is clear that in the main we are dealing with marriages not primarily regarded by those who enter into them as reproductive machines or devices whereby the husband may accumulate capital from a rich bride. If we wish to examine these functions, we have to turn to the fictive family of protégés, that multiplier of the patron's credit and earning capacity, not to the biological family.[33] How, then, are we to view these marriages? The usual historical approach toward marriage as a strategy of advantage is inappropriate in many cases.[34] Instead we have to look more carefully at attitudes toward marriage and at the kind of women who married professional savants, members of the Institut.

In collecting information on wives and marriage in this group, we encounter many different sorts of women and many different types of marriage. We need to examine these wives as a group to find out what sort of women they were, what social classes they originated in, what sort of bond they constituted between Institut members, and what other sections of the elite they linked them to.

No attempt has ever been made to produce a collective portrait of the wives of any sector of the Napoleonic elite, let alone of the scientific elite.[35] Partly this is a result of historians' tendency to produce a history in which only men appear as the protagonists and thus cannot offer a picture of their personal relations as in any way constitutive of their historical role. This failing has been taken to the opposite extreme by those who have written about women; they also tell their story without taking into account the marriages in which their protagonists passed the greater part of their lives. The problem in constituting a collective portrait of the wives is thus not primarily that of lack of information, for many, in particular nineteenth-century historians, contributed much to them; rather, it is one of establishing an approach to the evaluation of the importance of marriage in an elite whose marriage patterns cannot be subsumed under the analysis of strategies of advantage of the more obvious kind. The first positive point to make, however, is that a sizable proportion of Institut wives wrote novels, plays, pictures, translations, and political writings. And these were not merely ornamental exercises. Mme. Daubenton's novel, *Zélie dans le désert*, went through twenty-one editions after its first appearance in 1787. Mme. Biot published a translation of Ernst Fischer's standard work as *Physique mécanique*; it went into four editions. Mme. Fleurieu was the author of two comedies and a novel. From a preliminary list of similar achievements,[36] there emerges a group of women who to a large extent inhabit the same intellectual world as their husbands; they do not relegate themselves to a mysterious background world of children and household that has no impact on the world of the man. Marriage with such women meant a marriage of companionship rather than a marriage of biological reproduction.

From learning more about the Institut wives, another striking point emerges. One cannot fail to be struck by the extraordinarily disparate social origins involved, which range from Desfontaines's assistant gardener and Joseph Gay-Lussac's shop assistant, to Gaspar de Prony's, Pierre Charles Lemonnier's, and Pierre Leroy's marriages into the nobility, Jean-Baptiste Delambre's and Nicolas Deyeux's into the financial elite.[37] Historians of the Napoleonic elite have pointed out that elite status and income seemed to have little to do with each other in imperial France.[38] But the Institut, even disregarding disparities of income, does not show a stable class structure formed by men uniformly marrying "up" in a way that strengthens their claims to belonging to an elite. Rather, the Institut demonstrates an insouciance about class in the choice of a marriage partner that tended to dissociate prestige, wealth, and, as far as we can use this terminology today, social class.

Historians of ancien régime elites have frequently remarked on their tendency not to intermarry, despite financial interests in common or common membership of institutions.[39] When we turn to the Napoleonic Institut, however, we find that although not all members are related to each other by marriage, an option in any case ruled out by the high numbers of at least nominal bachelors, a substantial amount of intermarriage does occur. Men who married a former colleague's widow, or who were the brother-in-law of a colleague, or who married their daughter to a colleague – a tenable option when all elite groups were accustomed to marriage partners widely different in age – all are relatively common in our group[40] Such intermarriage cuts across subject boundaries embodied in the four classes of the Institut. This fits in with what we know about the lack of definition of discipline boundaries in this period.[41] Members of the Institut thus had little incentive to use marriage in a widespread way in order to increase their credentials within a well-defined scholarly group.

More interesting questions are raised when we consider possible reasons for the relative frequency of intermarriage among the scholarly elite of Napoleonic France compared to what we know of its ancien régime counterpart. We may speculate on the effects of increased pressure from the Napoleonic state on the Institut to create a greater feeling of identification among its members, which could be accomplished through increasing intermarriage. Or this could be one of the first stages in the growth of a self-conscious intelligentsia like that of post-1830 France.[42] More probably, however, increased intermarriage should be viewed as one of the first indications that the relatively undifferentiated elites of the Napoleonic state were gradually changing into the vertical elites of modern France.

This brings us naturally to a consideration of the extent to which marriage linked the members of the Institut to other sectors of the elite. This is not simply a measure of elite fluidity in the Napoleonic state as a whole, but also of the relative status of the Institut *in* that elite: rarely does a man willingly marry his daughter to someone he considers his social inferior, preferring rather to marry her to his equal or, even better, his superior. The question of where the intellectual elite as such stood in the pecking order of Napoleonic France, its perceived importance by other sectors of the elite, has rarely if ever been considered, even in social studies of such groups. Here, however, we encounter a difficult problem. Many members of the Institut performed multiple roles within the state; a few had wealth deriving from successful business enterprises.[43] How are we to establish whether such men were viewed by their prospective father-in-law, let alone by their prospective bride, as members first and foremost of the Institut, or as important bureaucrats or wealthy men who happened to have the right to wear an academician's uniform?

Put like this, the question is almost impossible to answer. We can approach it in two other ways, however: by pointing out that membership in the Institut, at the very least, does not seem to have deterred other sectors of the elite; and by using those members of the Institut who had no other occupation as a control group. There are no recorded examples of any member solely dependent on his vocation for income marrying "up" into nobility or higher bureaucratic families.[44] Institut membership by itself therefore conferred little chance of acceptance into the wider elite. But we must also note that membership in the Institut was usually connected with bureaucratic involvement of some kind. As one of the major new pools of talent in the Napoleonic state, a state in continuous expansion, bureaucratic employment was the normal fate of the intellectual elite.

I have argued that patronage in science cannot be fully understood without insisting on the role played by women. They not only provided the social setting in which future protégés could collect within the ambit of their patron, but also played a crucial role, one fully acknowledged by contemporaries, in the sifting and sorting of the patron's clientele and in the unfolding of individual consciousness of a scientific vocation. In an era before the full institutionalization of science, women, just as much as the male patron, provided by conscious exertion of social art and psychological insight a medium through which the aspiring young savant could locate his authentic self, the self his biological parents had refused to allow him.[45] That authentic selfhood was seen as the sine qua non of the ability to view the natural world correctly.

Women, dominant in the social world but largely excluded from formal decision-making structures, were capable of reinforcing authentic vocation precisely because their input was remote from the world of alliances, intrigue, and advantage in which the male patron, however unwillingly and ambivalently, was forced to make his way, a world seen as the enemy of personal authenticity. If patronage systems are means of cultural reproduction as much as of transmission of authority, then the system reproduced all the ambiguities of the scientific ideal. On the one hand, the otherworldly qualities of the true savant, located in authenticity, were enhanced by the women involved in the patronage network; on the other, the male patron provided entrée into *le monde*, the world that threatened personal corruption and the loss of the true self, and yet supplied the indispensable real world, means for the actual implementation of vocation.

Women in the scientific elite were able to perform these functions because of the specific demography of the group. They also performed them by virtue of a certain style of marriage, one whose function was seen as companionate activity rather than biological reproduction. Given the trend of much current research to identify marriage and career as hostile entities for aspiring professional women scientists from the mid-nineteenth century on, it is worthwhile to remember that in an earlier epoch, when no "professional" careers in science were given women and very few were open to men, the values and functions of marriage were able to mesh together relatively easily. Presented by many eighteenth-century thinkers as the most "natural" of the social institutions, wives and husbands, in their different ways, were able to heal the many breakings of bonds that characterized, in contemporary estimation, the typical career of the budding savant, and they were able to overcome the many contradictions in the scientific ethos.

Why was the position after mid-century so different? It cannot be wholly ascribed to the new problem of professional careers for women, or some idea of "necessary" conflict between career and family. It was much more that the nature of marriage itself changed after the impact of industrialization not only on economic life, but on value systems as well. Historians have often described these changes as an increasing tendency among the middle class to allocate the domestic, "intimate" interior to the wife, and at the same time to see the cultural role of "the home" (i.e., the wife's role in marriage) as a conservative rather than an innovative one. In setting up an automatic opposition between career and marriage we are also forgetting that the full professionalization of science was a long drawn out process, productive of insecurity at every level. It was not, after mid-century, that "all men" were preoccupied to deny "all women" their rightful place in the scientist's profession (although some clearly were); at a deeper, structural level, we can see that they were encouraged to make that exclusion more by changes in the nature and ideology of marriage, as well as by the continuing insecurity in science itself, both as ethos and as career.

Another consequence of industrialization was an increasing emphasis on ideals of objectivity, rarely posed as such in the science of the late eighteenth and early nineteenth centuries. However one defines objectivity, such an ideal has two clear consequences: to separate subject from object, observer from that which is observed; and to insist on the ejection of emotion from the processes of cognition. In doing so, it removed the ideological base from the important female input into the development of scientific vocation. If the inner world of the observer had no impact on the quality of his observing, the drawing out of individual self-awareness was of comparatively little importance, and science became far more the *doing* of a certain set of activities than the *being* of a certain sort of person, who could be helped to full authenticity by the emotional insights of women.

It was in this very different set of expectations that women were to struggle for careers in science after mid-century. A vision of women as complementary to science collapsed at the same time as did the culture that linked personal authenticity to the struggles of the savant.

Notes

1 A.E.M. Grétry, *La vérité, ou ce que nous fûmes, ce que nous sommes, ce que nous devrions être,* 3 vols. (Paris, 1807), 1:218; quoted in J. Simon, *Une académie sous la directoire* (Paris, 1885), 89.

2 See D. Outram, *Georges Cuvier: Vocation, Science and Authority in Post-Revolutionary France* (Dover, N.H., 1984), 169–202; idem, "Politics and Vocation: French Science, 1793–1830," *British Journal for the History of Science* 13 (1980):27–43.

3 For new work along these lines for a later period, see Françoise Mayeur, "Woman and Elites from the Nineteenth to the Twentieth Century," in *Elites in France: Origins, Reproduction and Power,* eds. J. Howorth and P. G. Cerny (London, 1981), 57–65. The importance of family groupings to cultural innovation and reproduction is also discussed in A.F.C. Wallace, *The Social Context of Innovation: Bureaucrats, Families and Heroes in the Early Industrial Revolution* (Princeton, N.J., 1982).

4 See Robert Mauzi, *L'idée de bonheur, dans la littérature et la pensée françaises au dix-huitième siècle* (Paris, 1960), 44–91, 147, 268–275, 355–485.

5 For technical philosophy supporting this view, see the theories of perception described in C. Van Duzer, *Contributions of the Idéologues to French Revolutionary Thought* (Baltimore, Md., 1935). The philosophy of Immanuel Kant only became influential in France after around 1800; see M. Vallois, *La formation de l'influence kantienne en France* (Paris, 1927); François Picavet, *La philosophie de Kant en France de 1773 à 1814* (Paris, 1888); K. Figlio, "Theories of Perception and the Physiology of Mind in the Late Eighteenth Century," *History of Science* 13 (1975):177–212. D. Outram, "The Language of Natural Power: The Funeral *Éloges* of Georges Cuvier," *History of Science* 16 (1978):153–178.

6 Ramond wrote in 1777: "Je regarde le marriage comme le lien le plus naturel, celui où la société a moins ajouté, celui dans lequel elle a le moins perverti le voeu de la nature, conséquemment celui dont le bonheur est le plus proche"; quoted in Cuthbert M. Girdlestone, *Poésie, politique, Pyrénées: L. Ramond 1755–1827: Sa vie, son oeuvre littéraire et politique* (Paris, 1968), 75.

7 D'Alembert wrote to Lagrange in 1767: "J'apprends que vous avez fait ce qu'entre nous, philosophes, on appelle 'le saut périlleux' "; quoted in Antoine Guillois, *La Marquise de Condorcet: sa famille, son salon, ses amis, 1764–1822* (Paris, 1897), 65.

8 Guillois, *Marquise,* 40–67.

9 Modern sociologists, unlike historians of science, take it for granted that such are the functions of family relations; e.g. P. Bourdieu, "Champ de pouvoir, champ intéllectuel et habitus de classe," *Scolie* (i) (1971):7–26; on p. 10: "Il faut se rappeler que parmi l'ensemble des privilèges qui sont l'instrument et le produit du pouvoir, il n'en est sans doute pas de plus important que le capital des relations; par l'intermédiaire du réseau de relations familiales et amicales s'opère un nombre important de transactions objectivement politiques et objectivement économiques."

10 Georges Cuvier, *Éloges historiques,* 3 vols. (Paris, 1819), 1:168: Éloge of Darcet. Such ruptures with the biological family as the price of personal authenticity had long been part of the formation of religious vocation; see Laurent Theis, "Saints sans famille? Quelques remarques sur la famille dans le monde franc à travers les sources hagiographiques," *Revue historique* 255 (1976):3–20; on p. 6: "Le saint n'est lui-même qu'après que sa famille a été annihilié éventuellement au profit d'un groupe familial métaphorique." See also S. M. Silverman, "Parental Loss and the Scientist," *Science Studies* (1974):259–264; W. R. Woodward, "Scientific Genius and the Loss of a Parent," *Science Studies* (1974):265–279.

11 Cf. Laurent Theis, "Saints sans famille," 6.

12 For further historical parallels, which point to the deep historical roots of scientific vocational ideals, see A. D. Nock, "Conversion and Adolescence," in *Essays on Religion and the Ancient World,* 2 vols. (London, 1972), 1:469–480.

13 Isidore Geoffroy St. Hilaire, *Vie, travaux, et doctrine scientifique d'Etienne Geoffroy St. Hilaire* (Paris, 1847), 39; other examples include Lacepède's remarks on Buffon: "Il me traîta comme son fils"; Roger Hahn, "L'autobiographie de Lacepède retrouvée," *Dix-huitième siècle* 7 (1975):49–85, esp. 58; Dupont de Nemours of Quesnay: "Je n'étais qu'un enfant lorsqu'il me tendait les bras. C'est lui qui m'a fait un homme," in H. A. Dupont de Nemours, ed., *L'enfance et la jeunesse de Dupont de Nemours* (Paris, 1906), 42.

14 Antoine Guillois, *Le salon de Mme. Helvétius: Cabanis et les idéologues* (Paris, 1894).

15 Charles Dupin, "Éloge de Mme. de Prony," *Mercure du dix-neuvième siècle* 2 (1822):203–215.

16 James F. Traer, *Marriage and the Family in Eighteenth-Century France* (Ithaca, N.Y., 1980); François Olivier Martin, *La crise du mariage dans la legislation intermédiaire, 1789–1804* (Paris, 1901); Roderick Philips, "Le divorce en France à la fin du dix-huitième siècle," *Annales* 24 (1979):252–287; J. Donzelot, *The Policing of Families* (New York, 1979).

17 E.g., Grétry, *Vérité*; J. A. Ségur, *Les femmes: leur condition, et leur influence dans l'ordre social,* 3 vols. (Paris, 1802), 3, 5–20; E. Legouvé, "La mérite des femmes," in *Oeuvres complètes,* 2 vols. (Paris, 1826), 2, 26–30.

18 Bonnie G. Smith, *Ladies of the Leisure Class: The Bourgeoises of Northern France in the Nineteenth Century* (Princeton, N.J., 1981), 4–51; Margaret H. Darrow, "French Noblewomen and the New Domesticity, 1750–1850," *Feminist Studies* 3 (1979):16–35.

19 J. B. Suard, "Éloge de Mme. d'Houdetot," *Journal des Débats* (February 6, 1813), 249. All translations from the French are my own, unless otherwise noted.

20 C. A. Lopez, *Mon Cher Papa: Benjamin Franklin and the Ladies of Paris* (New Haven, 1966), 243.

21 A. C. Kors, *D'Holbach's Coterie: An Enlightenment in Paris* (Princeton, N.J., 1976); Carolyn Lougée, *Le Paradis des Femmes: Women, Salons and Social Stratification in Seventeenth-Century France* (Princeton, N.J., 1976); Deborah Hertz, "Salonnières and Literary Women in Late Eighteenth-Century Berlin," *New German Critique* 1 (1978):97–108.

22 See the suggestions contained in D. Outram, "Politics and Vocation."

23 Dupin, "Éloge de Mme. de Prony," 213–214.

24 Henri Beyle, *Vie d'Henry Brulard,* ed. B. Didier (Paris, 1973), 380.

25 "Scientific elite" is a phrase that can only be used here anachronistically, since the period was distinguished by the social and intellectual interpenetration of different areas of intellectual inquiry, few autonomous disciplines were yet in existence, and many individuals practiced both "humanistic" and "scientific" disciplines with success. In statistical studies, I have established that the demographic behavior of the "scientific elite," which I define here with necessary crudeness as the membership and the members of the "First Class" of the Institut de France, does not vary significantly from that of the 272 individuals composing the Institut as a whole between 1795 and 1814.

26 Louis Bergeron and Guy Chaussinand-Nogaret, *Les masses de granit: cent mille notables du premier Empire* (Paris, 1979). Demographic information for the Institut is based on: P. Franqueville, *Le premier siècle de l'Institut de France,* 2 vols. (Paris, 1895); P. Gaujon, ed., *L'Académie des Sciences de l'Institut National de France* (Paris, 1934); *Institut de France: index biographique de l'Académie des Sciences du 22 décembre 1666 au 1 octobre 1978* (Paris, 1979); F. Michaud, ed., *Biographie universelle ancienne et moderne, nouvelle édition,* 45 vols. (Paris, 1843–48); C. C. Gillispie,

ed., *Dictionary of Scientific Biography,* 14 vols. (New York, 1970–78); J. Balteau, M. Barroux, and M. Prevost, eds., *Dictionnaire de biographie française* (Paris, 1932); M. de Courcelles, *Dictionnaire universelle de la noblesse de France,* 5 vols. (Paris, 1822); Léonce de Brotonne, *Tableau historique des pairs de France, 1789–1848* (Paris, 1889); François Michel, ed., *Fichier stendhalien,* 2 vols. (Boston, Mass., 1964); Alfred Potiquet, *L'Institut national de France, . . . 20 novembre 1795–19 novembre 1869* (Paris, 1871); Romuald Szramkiewicz, *Les régents et les censeurs de la Banque de France, nommés sous le consulat et l'Empire* (Geneva, 1974); Auguste Jal, *Dictionnaire critique de biographie et d'histoire* (Paris, 1867).

27 Bergeron and Chaussinand-Nogaret, *Masses de granit,* 14–17.

28 For example, Lamarck (*Dictionary of Scientific Biography*); Cuvier (D. Outram, *Georges Cuvier,* 65, 217); Bosc (Claude Perroud, "Le roman d'un Girondin: le naturaliste Bosc," *Revue du dix-huitième siècle* 2 [1916]:232–257, 348–367). See also R. Barroux ("Sebastien Mercier, le promeneur qui ne sait où il va," *Mercure de France* 64 [1960]:655–665); Jacques Payen (*Capital et machine à vapeur au dix-huitième siècle: les frères Perier et l'introduction en France de la machine à vapeur de Watt* [Paris, 1969], 56); S. Gillmor (*Coulomb* [Princeton, N.J., 1971], 75).

29 That is, those of Malus, Charles, Tessier, Daubenton, Delambre, Guyton de Morveau, Lacépède, Lagrange, Latreille, Lefèbvre-Ginau, Legendre, L. G. Lemonnier, Prony, Berthoud, Percy, Olivier (first marriage), Palisot, and Pinel (first marriage) produced no children. Marriages producing less than three children include those of Berthollet, Biot, Broussonnet, Cels, Desmarets, Deyeux, Girard, De Jussieu, Duhamel, Richard, Ramond, Silvestre, Deschamps, Mirbel, Messier, Buache, Coulomb, Defontaines, Fourcroy, Laplace, Geoffroy St. Hilaire, Carnot, Hallé, Monge, Bouvard, Claret de Fleurieu, Thenard, Ampère, Levêque, Adanson, Ventenat, and Pinel (second marriage).

30 For example, Bonnie G. Smith, *Ladies of the Leisure Class,* 5–6.

31 Jean Tudesq, *Les grands notables en France (1840–1849): étude historique d'une psychologie sociale,* 2 vols. (Paris, 1964), 1:458–459, 462–463.

32 For a specific example (the Brongniart family of naturalists), see D. Outram, *Georges Cuvier,* 174, 197–198; Louis Delaunay, *Une grande famille de savants: les Brongniart* (Paris, 1940).

33 For discussion of the role of the protégés, see Outram, *Georges Cuvier,* 189–202.

34 Such an approach would be typified by Pierre Bourdieu, "Les stratégies matrimoniales dans le système de réproduction," *Annales* 27 (1972):1105–1127. For a commercial élite, marriage for capital was inescapable; see P. Leuillot, "Bourgeois et bourgeoises," *Annales* 11 (1956):87–101.

35 This failure is common to all recent studies in this area; e.g., J. Houdaille, "Les déscendants des grands dignitaires du premier Empire au dix-neuvième siècle," *Population* 29 (1974):263–274; L. Bergeron and Guy Chaussinand-Nogaret, eds., *Grands notables du premier Empire,* 7 vols. (Paris, 1978–). Technical manuals pay little attention to the problem; e.g., Louis Henry, *Manuel de demographie historique* (Geneva, 1967).

36 A first listing from the Bibliothèque Nationale, *Catalogue des imprimés,* would include the following works produced by wives of men of science, members of the Institut: Mme. Biot (Françoise Gabrielle Brisson), *La physique mécanique* (1813; reprint, Paris, 1830), translated from the German of Ernst G. Fischer; Mme. Daubenton (Marguerite Daubenton, 1720–1818), *Zélie dans le desert,* 2 vols. (Paris, 1787), and twenty-one subsequent editions, 1787–1861; Mme. Claret de Fleurieu (Aglae Deslacs d'Arcambal, 1776–1828), *Au Théàtre de la Nation: le siècle des ballons, satyre, nouvelle comédie*

en 1 acte (Paris, 1784); *Stella, histoire anglaise,* 4 vols. (Paris, 1800). Mme. Guyton de Morveau (Claudine Poullet, 1735–1821), translations of Scheele and Stahl; Mme. de Lacépède (Anne Caroline Gauthier, née Jube [?–1801]), *Sophie ou mémoires d'une jeune religieuse* (Paris, 1790; 2d ed., 1792); Mme. de Lefrançois de Lalande (Marie Jeanne Emilie Harlay), *Tables horaires,* in J. J. Lefrançois de Lalande, *Abrégé de navigation* (Paris, 1793); Mme. Laplace (Marie-Anne Charlotte de Courty de Romanges [?–1862]), *Lettres de Mme de Laplace à Elisa Napoléon, Princesse de Lucques et de Piombino,* ed. P. Marmottan (Paris, 1897).

37 For Delambre and Deyeux, Szramkiewicz, *Régents et les censeurs,* 38, 268, 272; for Gay-Lussac, M.P.C. Crosland, *Gay-Lussac* (Cambridge, 1982); for Prony, Dupin, "Éloge de Mme. de Prony"; for Lemmonier, Michel Robida, *Les bourgeois de Paris: trois siècles de chronique familiale de 1675 à nos jours* (Paris, 1955), 45–46; for Leroy, G. Brusa and C. Allix, "Julien and Pierre LeRoy: Their Business, Their Relatives, and Their Namesakes," *Antiquarian Horology* 7 (1972):598–606; Lopez, *Mon Cher Papa,* 213–215.

38 Jean Tulard, "Problèmes sociaux de la France impériale," *Revue d'histoire moderne et contemporaine* 18 (1970):639–663.

39 For example, Yves Durand, *Finance et mécenat: les fermiers-généraux au dix-huitième siècle* (Paris, 1976), 165.

40 For example, in 1792 the mathematician P. L. Lagrange married the daughter (Renée Françoise) of the botanist Pierre Charles Lemonnier (1715–1799), whose third daughter, Renée Michelle, married her uncle the horticulturalist Louis Guillaume Lemonnier in 1794. In 1800 the chemist A. L. Fourcroy married Adelaide née Belleville, widow of Institut member architect Charles de Wailly. See Robida, *Les bourgeois de Paris;* for Fourcroy, G. Kersaint, "Antoine François Fourcroy, 1755–1809: sa vie et son oeuvre," *Mémoires du Muséum National d'Histoire Naturelle,* sér. D, 2 (1966):1–296.

41 J. Schiller, "Physiology's Struggle for Independence in the First Half of the Nineteenth Century," *History of Science* 7 (1968):64–89; D. Outram, *Georges Cuvier,* 118–140; idem, "Uncertain Legislator: Georges Cuvier's Laws of Thought in Their Intellectual Context," *Journal of the History of Biology* 19 (1986), 323–368.

42 See the suggestions in D. Outram, "Politics and Vocation."

43 Wealth from, e.g., chemical industry was possessed by Chaptal: Roland Peigeire, *La vie et l'oeuvre de J. C. Chaptal* (Paris, 1934). Multiple administrative positions are detailed in D. Outram, "Politics and Vocation."

44 An example here, passed to us with an unusual wealth of financial detail, is Paul Cottin's edition of the journal of Mme. Moitte, wife of sculptor Jean Guillaume (1746–1810); P. Cottin, ed., *Journal de Mme. Moitte* (Paris, 1932). The family only married into other "working" artistic families, in spite of Moitte's membership in the Institut from 1795.

45 For a detailed working out of family politics within one scientific group, see D. Outram, *Georges Cuvier,* 161–188.

'LE LANGAGE MÂLE DE LA VERTU'

Women and the discourse of
the French Revolution

From: Peter Burke and Roy Porter (eds.), *The Social History of Language* (Cambridge and New York: Cambridge University Press, 1987), 120–135. © Cambridge University Press. Reprinted with permission.

The history of the French Revolution used to be dominated by studies of class struggle, *crises de subsistances*, war and terror. But it is now increasingly recognised that it is impossible to understand the revolutionary phenomenon itself without examining the very special political discourse which it generated.[1] For François Furet in particular, that discourse is the central motor of the Revolution:

> Since the people alone had the right to govern – or at least, when it could not do so, to reassert public authority continually – power was in the hands of those who spoke for the people. Therefore, not only did that power reside in the word, for the word being public, was the means of unmasking forces that hoped to remain hidden and were thus nefarious; but also power was always at stake in the conflict between words, for power could only be appropriated through them, and so they had to compete for the conquest of that evanescent yet primordial entity, the people's will. The Revolution replaced the conflict of interests for power with a competition of discourses for the appropriation of legitimacy. . . . Revolutionary activity *par excellence* was the production of a maximalist language through the intermediary of unanimous assemblies mythically endowed with the general will. In that respect, the history of the Revolution is marked throughout by a fundamental dichotomy. The deputies made laws in the name of the people, whom they were presumed to *represent*; but the members of the *sections* and of the clubs acted as the *embodiment* of the people, as vigilant sentinels, duty-bound to track down and denounce any discrepancy between action and values, and to reinstate the body politic at every moment . . . the salient feature of the period between May–June 1789 and 9 thermidor 1794, was not the conflict between Revolution and counter-revolution, but the struggle

DOI: 10.4324/9781003038085-7

between the representatives of the successive assemblies and the club militants for the dominant symbolic position, the people's will.[2]

The nature of revolutionary politics, in other words, made political discourse central, and political discourse shaped the very motor of revolution itself. The Revolution was the first point in French history where persuasion of a mass audience was crucial and an integral part of the political phenomenon. Words, as Furet argues, were power. At the same time, the collapse not only of the institutions of the Old Regime, but also of its ideological legitimation, meant that a desperate need existed to create a new discourse of validation for the new State, and for the groups which competed for this control. Control of the discourse of the Revolution gave access to opinion, to the general will, and conferred the power which came from successful representation of that general will to itself.

We need to explain further why this discourse was capable of playing this role. The answer comes from the Revolution's rejecting and contorted attitude to power. Many writers have stressed the extent to which, in this discourse, power as such was seen as evil and contaminating, in other words was approached almost entirely in moral terms. Politics and morality were completely conflated.[3] Under the Revolution, claims to power thus came from the denunciation of power; but now, the denunciation of power had to be carried out in the name of a pure and undivided general will. Thus, what the Revolution *used* as its political discourse does not fulfil what Pocock has described as the minimum criterion that a discourse *be* regarded as political, that it should 'consist in the utterance of essentially disputed propositions'.[4] It contained no means of admitting that political morality might be variable, and it contained no way of discussing conflicting sectional interests which undermined its ideal of the endless, unbroken harmony of the general will, from which all revolutionary legitimacy stemmed. As Roger Barny has put it, 'Les théories politiques bourgeoises . . . évacuait les rapports sociaux concrets . . . [et] ne pouvaient pas se plier aux nécessités d'une lutte politique exprimant, dans son fond, la lutte des classes.'[5] Thus the discourse of the Revolution was an object of competition because it offered the false, though convincing, promise that it was not a political discourse at all, and thus that it guaranteed the purity and hence the legitimacy of its users.

How is this revolutionary discourse to be recognised? Firstly, by its ritualised invocation of absolute, moral concepts, of which the most important was that of 'virtue', as the guarantors of the integrity of the Revolution. Such concepts, like the 'general will', are hypostasized entities, hardly ever receiving described real embodiment; for if the nation is as one in its general will, and that general will is the guarantor of the legitimacy of the Revolution, sectional interests do not exist, and there are therefore no precise descriptive words for them. It is not surprising that this discourse is also full of human reference figures turned into embodied universals. To take only the most famous – and for our purposes, most significant – figure, that of the elder Brutus, we find that the political discourse of this period abounds in shorthand references to Brutus as the embodiment of virtue, a man who put the safety of the state above private emotion and the destruction of his family: 'virtue' and 'Brutus' are virtually reciprocal referents.[6] This is also a discourse which proceeds by the ritual invocation of polarities: vice and virtue, aristocrat and people. This is the reason that the exclusionary power of this discourse is so strong. He who is not with us is against us. And this is why struggle for appropriation of this discourse could literally take on a life and death character. Furet is surely correct to see this form of political discourse, which leaves no middle ground open for debate and negotiation, as one of the explanations of the high turnover of ruling groups in revolutionary politics.[7]

But we can take the description of revolutionary discourse further than Furet does, and ask other questions about it. First of all it is clear that in spite of its maximalist claims to embody the people's will, that discourse never replaced many other sorts of discourse which were used, when convenient, for political ends. Many of the key words of the discourse of the Revolution, such as 'virtue', also had powerful resonances in completely other discourses, and in particular in that discourse on femininity to which the eighteenth century had added so much elaboration. Such facts are crucial in women's encounter with the public language of the Revolution. In particular, the conflation of morality and politics within the discourse of the Revolution made the barriers between that, and the discourse on femininity, a highly permeable one.

We can also ask of this discourse, as we would of any other, where are its weaknesses and stresses? Furet's account shows us a monolithic discourse terrifying in its capacity to aid and initiate proscription and Terror. Yet it was also a discourse able, if necessary, to abandon its circular, self-consistent character, and ruthlessly to appropriate other discourses to itself. It was also a treacherous discourse. It posed more problems to the majority of its users than Furet admits, for there is no discourse which causes so many problems to its users as that one whose control is a life and death matter, which is itself a validation of the power to speak at all, but which is useless for the negotiation of their sectional interests. In the case of the bourgeoisie, we may concur with Patrice Higonnet's argument that this characteristic of the discourse accounts for the incoherence of bourgeois aims in the Revolution up to 1794, and for the length of time which it took them to gain unchallenged control of the revolutionary movement. In other words, to pursue sectional interests through a universalistic rhetoric may be necessary, but it is also inefficient and dangerous, even for social classes which *have* actual control of the political process, and even more for excluded groups. This is a problem which affects women's use of the 'revolutionary discourse' to the highest degree.

Women's use of and response to the discourse of the Revolution, and male reactions to their attempt to appropriate that discourse, is a topic which broadens our perspective on the nature and uses of political language. As an excluded sectional group *par excellence* women had to struggle for their own interests through the medium of a political discourse unsuited to their needs, as it was to the needs of every other excluded sectional group. But surprisingly enough even the simultaneous recent increase of interest both in the language of the Revolution and in women's participation in the Revolution has not resulted in any study of women's reaction to that discourse. Public utterances by women using the 'revolutionary discourse' have been the subject of great attention from feminist historians, but they have almost all concentrated on the message, the explicit content of their utterances, to the exclusion of the problems posed for women by the nature of the discourse itself. This partly stems from the Whig orientation of much of women's history, which leads it to concentrate on demands which can be seen as foreshadowing later and more successful 'feminism'. This approach leaves on one side the entire problem of the structure of public utterance itself, and what it tells us about the position of women in relation to the public sphere. If we knew more about that, then we might also be able to say more about the reasons for women's failure to achieve demands for political equality in this period.[8]

Feminist historians have also paid very little attention to the non-public relations of women in this period. Autobiographies, private letters and private diaries rarely figure in their accounts of the female response in this period, and with this omission we lose an essential alternative register to the political, private voices which were often as wrapped up in the revolutionary discourse as any man's. Once again, this failure to utilise private documentation reflects a tendency to stay within the canon of 'founding mothers', which, as has been frequently remarked, forms at once one of the strengths and the weakness of much women's history. It is certain, however, that concentration on continuous retelling of the acts of a small band of female activists such as Théroigne de Méricourt means that women's history in the revolutionary period may well continue to be curiously isolated from that of the general history of the Revolution, and that mainstream history will accordingly take little account of women's history. It is significant both of this isolation, and of this lack of sensitivity to the problem of women's response to the revolutionary discourse, that, for example, Régine Robin's analysis of public utterances by revolutionary leaders should not only ignore the whole issue, but also tacitly exclude public utterances by women from her analysis.[9] I would argue here for the use of a far wider selection of material produced by women than has usually been the case, for it is largely in informal, private writing that explicit comments on the use of language in public surroundings seem to occur. There is also the methodological point that far more interesting sorts of queries can be answered about language if different sorts of discourse can be identified and their uses contrasted with each other. The success of Paul Fussell's analysis of stresses in 'official' discourse through their reflections in such private discourses as soldiers' letters home and private diaries, is a continuous reminder of how the historian of discourse can proceed only if he has at his disposal that *variety* of registers which Furet has christened the 'dual keyboard' of utterance. Only thus can the historian become aware of the rules governing the use of different types of discourse, of their permeability to each other, and the tensions between them.[10]

The invisibility of the problem of women's encounter with public discourse is also encouraged by the state of contemporary sociolinguistics, which seems determined to provide as few models as possible to the historian. Contemporary sociolinguistics commits the reverse sin from that of the feminist historians of the Revolution, and tells us little if at all about women's speech outside the private realm, the home, and little too about how women talk to each other. In other words, it replicates the still persistent myth of women as private beings, creatures of an interior. There is as far as I know, no study of women's use of public discourses, let alone of the problems they might encounter in such use.[11] What follows can only be a tentative attack on the problem, but it is clear that this topic would tell us not only a great deal about women's political difficulties under the Revolution; it would also tell us much about the stresses within that revolutionary discourse of which Furet has given us so monolithic a picture.

Problems in the use of a specific discourse may also arise not simply from the inner character of that discourse, but also from the way it is socially inserted, its resonances against the rest of speech. Revolutionary discourse was not sex-neutral. Its main emotive words, such as 'virtue', were ones which had a long history behind them, not only in the tradition of civic humanism, but also of weighing more heavily on the female sex than on the male. All sectional groups were disadvantaged by this discourse, but at least the dispossessed, the 'people' at large, could claim to be the general will, *le souverain*, and thus to embody political virtue. The fact that it was difficult to identify the sovereign and its will, and that the bourgeois leaders of the Revolution assumed control of that definition for their own purposes, should not lead us to overlook the fact that in the interior of that discourse *le souverain* was automatically Good.[12] Women as such, in this bi-polar world of reference, were Bad. To a very large extent, the influence of women was seen as the defining characteristic of the corruption of power under the Old Regime. Boudoir politics, the exchange of political gifts for sexual favours, are seen both as causes of the weakness of the Old Regime, and as the justification of the Revolution. Perhaps the most telling example of this occurred during the trial of Marie Antoinette, the deposed Queen of France, in 1793. The 'political' counts against her – of aiding the King's flight to Varennes, of plotting the invasion of France – were inseparable from, and bolstered by, the accusations of sexual perversion and incest which accompanied them. In 'corrupting' the Dauphin, the heir to France, with sexuality, her accusers implied that she had corrupted the body politic at one and the same time as she had corrupted the actual physical body of her son.[13]

As Furet remarks, the Revolution, and with it, its particular discourse, could only legitimate its own capacity to wield power, and distinguish itself from the corrupt power of the Old Regime, by attacking power itself.[14] To the degree that power in the Old Regime was ascribed to women, that meant that the discourse of the Revolution was committed to an anti-feminine rhetoric. It was a rhetoric with great investment in it for the male politician, and analysis of it from this direction may explain many of its salient features in a way not attempted before. In the analysis of the old regime enormous power was ascribed to women – an ascription which functioned to absolve men from responsibility both for the weakness of the monarchy and for its overthrow. Male politicians could find in this analysis an escape from the guilt necessarily arising from their participation in the destruction of an entire order of society and its complex religious sanctions. What looked like a sacrilegious act had *in fact* been a crusade for virtue. It is more relevant for our purposes, however, to note how this analysis of the sexual corruption of the Old Regime, and its legitimation of the Revolution, turn on a slide in meaning between two definitions of 'virtue'. Women's personal virtue (virtue = chastity) is equated with political virtue (virtue = putting state above personal or sectional interests), like Brutus, who executed his sons when they attempted to betray the Roman republic. The continuum between the two senses carries a whole series of messages: that female chastity is the prerequisite for political innovation undertaken in the name of a universal will; that women are in any case threatening to the Revolution, because any deviation from chastity = virtue involves the collapse of political virtue, and because women can personalise politics with their sexual favours, and factionalise it with competition for those favours. 'Virtue' was in fact a two-edged word, which bisected the apparently universalistic terminology of *le souverain* into two distinct political destinies, one male and the other female. All citizens were part of *le souverain*, but somehow one half of *le souverain* could only function at the price of the sexual containment of the other half.

It was through the word 'virtue', so crucial to the discourse of the Revolution, that power was taken from women both in the interior of that discourse, and as users of that discourse. It is also no accident that many of stock adjectives of the revolutionary discourse – 'chaste vertu', 'probité austère', 'vertu mâle et républicaine' – should enforce this impression of a discourse heavily weighted in gender terms. The accepted historiography of the Old Regime ascribed enormous power to women, power partly defined as sexual evil. The necessary conclusion was that political revolution could only take place if women were excluded from exercising power, and the niche formerly occupied by women's powerful vice was taken over by male virtue. As Olympe de Gouges remarked, 'women are now respected and excluded, under the old regime they were despised and powerful'.[15] All recent writing on women and the Revolution, which seems to emphasise women's *power*, however temporary, must explain why contemporary perceptions were so radically different.

And so it came about that the woman who refused to be respected but excluded was faced with a series of difficult choices, in the actual act of speaking in public as much as in the choice of words to speak. If she was not respected, there was no way of being heard; if heard, then the way was open for all sorts of attacks on a consequent loss of virtue – just the sort of attacks that were in fact made on the female activists of the political clubs and especially on the women's political club, the Société des républicaines révolutionnaires.[16] This explains some of the apparently very contorted attitudes taken up by women who wanted both participation *and* respect. One method was that provided by Mme Roland: to provide a forum for others, men, to speak in, whilst remaining silent, and hence respected, herself. Few more frustrating ways of becoming one of the best-known figures of the Revolution can be imagined:

> Je savais quel rôle convenait à mon sexe et je ne le quittai jamais. Les conférences se tenaient en ma présence sans que j'y prisse aucun part; placée hors du cercle et près d'une table, je travaillais des mains, ou faisais les lettres, tandis qu'on déliberait . . . je ne perdit pas un mot de ce qui se débitait et il m'arrivait de me mordre les lèvres pour ne pas dire le mien . . .
>
> Se taire quand on est seule n'est pas chose merveilleuse; mais garder constamment le silence au milieu des gens qui parlent des objets auxquels on s'intéresse, réprimer les saillies du sentiment qui vous oppriment lors d'une contradiction, arrêter les idées intermédiaires qui échappent aux raissonneurs, et faute desquelles ils concluent mal ou ne sont pas entendus, mesurer ainsi la logique de chacun en se commandant toujours soi-même, est un grand moyen d'acquérir de la pénétration, de la rectitude, de perfectionner son intelligence et d'augmenter la force de son âme.

Mme Roland's continual emphasis in her *mémoires* on her private, domestic, solitary life, even during Roland's two ministries, comes straight out of her acceptance of the idea that women can be valued only inasmuch as they wield corrupt power: 'on n'avait cherché à me voir que dans l'idée qu'il pouvait en être dans l'ancien régime, où l'on engageait les femmes à solliciter leurs maris'. When she was finally arrested and imprisoned in St Pélagie, she found with a degree of shock explicable only on these lines, that the authorities had taken little care to segregate her from the ladies of easy virtue (revealing phrase) also held within its walls: 'Voilà donc le séjour qui était réservé à la digne épouse d'un homme de bien. Si c'est là le prix de la vertu sur la terre, qu'on ne s'étonne donc plus de mon mépris pour la vie.' The long fight to possess both virtue and participation seemed to have passed so unsuccessfully that she was no differently viewed from the ladies of the town. It is against limitations like these, imposed by revolutionary discourse, that we have to redefine the *sort* of power Mme Roland in fact possessed.[17]

The conditions set for women's use of the revolutionary discourse by male politicians similarly contained enormous internal contradictions. It seemed that it was only within the sphere of sexual containment *par excellence*, the married home, that women could use the language of the revolution with the blessing of male political leaders. As a *section* leader from Orleans remarked in 1792: 'Les mères approndront à leurs enfants à parler de bonne heure le langage mâle de la liberté.'[18] All very well. But what to think of that language – and hence of the viability of the task of political education entrusted to women – when it was itself highly hostile to that guarantor of female respectability, the home? As remarked before, revolutionary discourse gloried in personified abstractions of virtue, most of them culled from the more austere, inflexible and unlovable figures of the history of the Roman republic. Brutus, a favourite reference figure to the men, was a source of especial difficulty to the women. Many women, in fact, found the entire Roman reference point difficult to sustain. A reading of private sources shows how many women, of all political views, and including many who consciously accepted their relegation to the domestic sphere, none the less reject Rome. Mme Jullien, wife of the significantly named Marc-Antoine Jullien, Jacobin deputy for the Drôme, went through a phase of very conscious rejection of the whole 'Roman fever', as she put it: 'Il faut redescendre au niveau commun, et penser que le mieux est l'ennemi du bien. Je me suis donc guérie de ma fièvre romaine, qui pourtant ne m'a jamais fait donner dans le républicanisme que par la crainte de la guerre civile.' Mme Roland's reactions were similar; and Mme Cavaignac, republican and mother of the future President, brought out the ambiguities better than anyone when considering the problem of teaching about republican virtue through the figure of Brutus:

> Il ne me paraît pas prouvé que Cataline lui-même fut le Cartouche qu'on nous a montré, et les injures de Cicéron, ce type de modérantisme, ne m'en donnait pas la certitude . . . Hélas, on ne se tromperait guère en ce monde, surtout dans ce bon temps de *chaste verité*, en tenant toujours pour prouvé le contraire de ce qui est officiellement établi . . . [for a brief period, however, even she was misled, and] . . . je tenais les deux Brutus en grande vénération: le premier qui a tué son fils, le second qui a tué son père.[19]

It was not simply that there was a 'separate sphere' for the exercise of female political virtue – it was that male personifications of virtue actually attacked and destroyed what they themselves defined as the physical location and necessary condition of female virtue, the home and family. For a woman to speak this indeed male language was in effect to endorse the destruction of her sphere, the home, should the higher male definition of virtue demand it. This would leave a woman, in the last analysis, with precisely nothing: neither a secure sphere of discourse for herself, nor an easy access to that of men.[20] Such an inconsistency could do nothing but undermine the authority of any woman who tried to use this revolutionary discourse, and it undermined it in ways never experienced by the male *sans-culottes*.

Authority in political discourse, in this time of complete conflation of the moral and the political, came from nothing so much as a perceived congruence between the moral discourse implicit in the words, and the capacity for moral personification which rested in the figure of the speaker. This was the source of Robespierre's power, as Furet has pointed out: that he fully personified the discourse he used.[21] He thus demonstrated the revolutionary discourse's capacity to legitimate not only a political order but also a *person*, by abolishing the incongruence between public and private roles. Women, on the other hand, were dubious about the enterprise of personification. Because their revolutionary respectability rested on their sexual containment within the male-dominated world of marriage and family they could never personify themselves as whole beings within that discourse. It is perhaps significant that the only group of women permitted to make a direct input into revolutionary discourse were actresses. Many a revolutionary public figure went to the great ladies of the Comédie Française for instruction on how to be, how to personify themselves.[22] But actresses were in any case not ordinary women: their profession turned them into embodiments of pure personification. Other, ordinary women, tended to view the enterprise with suspicion. Mme Cavaignac, for example, thought that her mother had too grand and simple a personality to *need* to be given a Roman name, such as Cornelia, mother of the Gracchi.[23] In refusing personification, women were also perhaps refusing to participate in the first political discourse of modern times to have that defining characteristic of modern 'totalitarian languages', in Jean Pierre Faye's phrase, which is the attack on the family.[24]

Here, the discourse of the Revolution is, of course, enormously different from that of the Old Regime, whose basic predicate was that of the intimate human body (the body politic), and the intimate family (the King as father of his people).[25] The discourse of the Old Regime is hesitant precisely in its relations with *le bien public*; that of the Revolution on the contrary is confident about the general will, but wishes to know nothing of the family in society, or the family *as* society.

We can thus also say that it may be no accident that a far more normal female response to the Revolution was not the contorted encounter with its discourse that we have so far been describing in the cases of women strongly tied to the new movement through family links and personal conviction; it was, far more, the rejection of the Revolution itself, and of the discourse that went with it. It has long been a commonplace of the history of the Revolution, from Michelet on, that counter-revolution was a movement in which women played an important, and enduring role, far more than they did in the Revolution itself. Recent research – as so often – has done nothing but confirm Michelet's perception.[26] Yet this is a fact that goes virtually unrecognised in the 'women's' history of the Revolution, which understandably enough prefers to focus on an image of woman as a political activist, and an activist on the correct, left-wing, progressive side. Yet it is now becoming clear that women's devotion to throne and altar during the Revolution were to set the stage for that decisive separation between men's politics and women's politics which occurred in the course of the following century, and led to the formation of radically different attitudes between the sexes on such crucial issues as the fate of the Church and of republicanism.[27]

139

It is also rarely, if ever, remarked that the Church had the supreme advantage, from a woman's point of view, of possessing the only other universalistic discourse on offer in the revolutionary period – and certainly the only other discourse with anything like the resonance of the discourse of the Revolution. It is this language which is overwhelmingly used by ordinary women who decide to protest against aspects of the Revolution. It was the language of the Vendée. It is also used even within the highly politicised working class of Paris itself. The women, of whom little is heard in current feminist history, who did not endorse the *sansculotte* demands of bread and Terror, overwhelmingly use the languages of the Gospel in their appeals for mercy.[28] Claude-Françoise Loissillier protested against the guillotine, for example, in the following way: 'C'est attaquer tout à la fois le Créateur et la créature: le Créateur, en détruisant son ouvrage; la créature, en la privant du bienfait de Dieu. Craignez surtout que cela n'attire sur vous et sur cette grande ville les grands fléaux de Dieu, en laissant faire cela plus longtemps.' Melanie Ernouf protested against the fall of the monarchy in very similar language: 'J'aime mon roi, je le regrette tous les jours et veux le suivre et me jetter dans les mains de ces vils assassins. Ils aiment les victimes: qu'ils s'abreuvent du sang pur des agneaux'. This discourse also competes with the 'revolutionary discourse' at the highest levels. It was long ago pointed out by Albert Mathiez, that many of the female prophets who emerge during the Revolution had contacts with male politicians which seem to have been more numerous and less obviously hostile than those which characterised the more 'feminist' women.[29]

So far, we have simply considered the many and considerable disadvantages to the female speaker of the discourse of the Revolution. Yet in spite of this very negative analysis we are still faced with the problem that women did use this discourse repeatedly in their public statements, and that they used it in public contexts even in utterances made very much in the heat of the moment, as Soboul has pointed out.[30] Citizeness Auxerre, silk-weavers' leader, in the heat of the *prairial* days of 1795 burst out with, 'qu'ils étaient le souvereign, que les officiers municipaux et les autorités n'étaient que leur agents . . . et qu'il était bien étonnonant que le souvereign manquait de bois quand ses agents en étaient abondamment pourvus'. So we cannot, and do not, mean to establish the discourse of the Revolution simply as yet another historical obstacle for women in their efforts to gain political equality with men, having nothing to do with their real inner selves. Some reasons why women might *want* to use this discourse in public statements are obvious enough; their audiences, except in the very unusual case of the Société des femmes républicaines révolutionnaires, were generally of both sexes. It makes sense, further, for an excluded or peripheral group to use the discourse of dominant groups in demanding change, in order to gain the attention of such groups and to validate their own utterances. But there is another sense in which any universalistic language, whether it be the discourse of the Revolution, or of Christianity, has a positive function for women, as well as the very negative features we have so far concentrated upon. In one of the few studies in modern sociolinguistics to contribute much to the problems of women's historical encounters with public language, Robin Lakoff has pointed out that in 'ordinary life' women are continuously confronted with an impossible choice of language; if a woman

> refuses to 'talk like a lady', she is ridiculed and subjected to criticism as unfeminine; if she does learn, she is ridiculed as being unable to think clearly, unable to take part in a serious discussion: in some sense, as less than fully human. These two choices a woman has – to be less than a woman, or less than a person – are highly painful.[31]

There is no way out of this linguistic catch-22 (one which, as we have seen, trapped Mme Roland completely). But discourses which *appear* to be univer-salistic can be seen as a way of avoiding the problem for the female speaker, because they often offer an automatic authority and validation. There is also the point that the problem of 'lady-like language', arises and is enforced in very specific social situations; whereas the promise of the discourse of the Revolu-tion, with its very rigidity, its set phraseology, its routinised allusions to the great personifications, is that it is relatively *impermeable* to the precise social and public setting in which it happens to be used. For the male speakers in the National Convention, for example, it was positively a disadvantage to be caught within the confines of the discourse when facing not only an audience of their peers, but also the less calculable reactions of the Parisian mob which packed the public galleries. But for women, it may well have seemed to work the other way, and the discourse's very rigidity may have offered them escape from the perennial moulding of the social situation which insisted on their femininity as the reason for denying them the right to speak with authority as public per-sons. The fact that all these hopes invested in the use of the revolutionary dis-course by women were in the end disappointed, and that so many women came to reject the discourse itself, does not invalidate these points about the initial attractiveness of the language to the female speaker. This means that the relation of women to the revolutionary discourse is ambivalent and close, rather than one of complete rejection. In spite of its appearance, and in spite of the promises of the Revolution itself for universal citizenship, the discourse of the Revolu-tion was far from being sex-neutral. But it was an instrument that women used in great numbers, whatever the misgivings and criticisms that many of them expressed, because of the way it seemed to promise authority to the otherwise compromised position of the female speaker in a public arena.

How has this necessarily brief survey helped us to understand more about the problems of women, and the nature of that discourse? To take the last point first: it is clear that certainly in the work of commentators such as Soboul, and less explicitly in that of non-Marxist historians such as Furet, linguistic analysis is dominated by class analysis. Revolutionary discourse is seen pri-marily as an exercise in mystification, a means whereby bourgeois politicians

prevented the excluded social and economic groups from producing an autonomous discourse which would have given names to their specific discontents.
The mystificatory use of language for the purposes of class warfare was not
lost on contemporaries either. As the Abbé D'Olivier remarked, 'l'abstraction
des droits égaux est une illusion redoubtable, et sans doute une supercherie
inventée par les puissants et les riches'; the popular leader, Jacques Roux, took
up the point two years later: '"Liberty" is naught but an empty dream if one
class of men can starve another with impunity. "Equality" is naught but an
empty dream if the rich through their monopolies can exercise the power of
life and death over others.'[32] So class in this view dominates language, and
gives it its hidden relationships of power. This is a view with much truth to it.
But once one asks the question of how the revolutionary discourse affected a
gender rather than a social class, then whole new functions of this language
are revealed. It becomes obvious that it is inadequate to analyse the discourse
of the Revolution only as a more or less consciously utilised tool of class war;
it is also, and just as much, a language fashioned for the war of the sexes.[33]
In fact one could argue that the whole of the discourse, in its choice of adjectives, in the weight it puts on the gender-laden concept of *virtue*, in its whole
implicit and explicit glorification of woman's sexual containment, in its identification of the female as all-powerful and all-corrupting, capable of changing
the entire nature of political regimes by the use or possession of her body, is
even more orientated to the mystification of women than it was towards the
mystification of excluded social groups. Instead of the lynch-pin of the whole
discourse being the word *virtue*, that word in fact stands revealed as the point
of *maximum stress* within that discourse, and the point at which that discourse
is most affected by the reverberations of its terminology in other, more 'normal' discourses. Women's position here challenged the gap between the reality
of politics and the discourse of politics even more than did that of the socially
and politically excluded as such, groups to whom the discourse automatically
attached *virtue* from within the centre of the discourse. In the case of these
excluded groups, the concept of virtue was of major importance in 'evacuating'
(to use Barny's phrase) the discourse of a concrete political, factional reality;
in the case of women, it was used not so much to evacuate political reality as
to divide it and both to stigmatise and confine the female half of it. As long
as women left unchallenged the equation of sexuality with political corruption, there was no way that they could arrogate to themselves and their cause
the magic, validating power of an unproblematic use of the central concept of
revolutionary discourse, that of virtue. This sentiment was even concurred in
by the 'feminist' activist Olympe de Gouges:

> Women have done more harm than good. Constraint and dissimulation
> have been their lot. What force robbed them of, ruse returned to them;
> they had recourse to all the resources of their charms, and the most
> irreproachable person did not resist them. Poison and the sword were

142

both subject to them . . . The French government especially, depended throughout the centuries on the nocturnal administration of women; the cabinet kept no secret from their indiscretion . . . anything which characterised the folly of men, profane and sacred, all have been subject to the cupidity and ambition of this sex . . .

The revolutionary tribunal agreed, and sentenced exemplary numbers of prostitutes to death as a moral *and therefore political* menace:

Le despotisme a toujours été l'ennemi des mœurs publics, que la prostitution était un des moyens qu'il employait pour affirmir son empire et perpétuer l'esclavage des citoyens par l'appat du libertinage et de la débauche; qu'on ne peut plus douter que les répaires de prostitution ne soient les asiles ordinaires des contre-révolutionnaires qui payent leurs infâmes plaisirs avec l'or de Pitt.[34]

It is thus obvious, to go to our original point, that one of the main explanations of why women, like the urban working class, found so little permanent gain from the revolutionary period may be linguistic. The series of deceptions and catch-22 positions in which the discourse of the Revolution ensnared them contributed enormously to their lack of long-term public authority. The failure of a 'women's movement' using a male-orientated political discourse was not only, however, completely foreseeable: it had the converse consequences, that those women who campaigned against the Revolution, using another universalistic discourse, far more adapted for women's needs as public persons, that provided by the Church, were those who were going to set the trend of women's political role, for all except a minority, for the next century in France. That role would be founded on a total separation of the political ideologies of men and of women, leaving men to reject the programmes and language of the Church and women to reject the programmes and language of secular republicanism.[35] Arguably, therefore, the discourse of the Revolution succeeded perfectly in carrying out its 'hidden agenda' of the exclusion of women from a public role, leaving for them recourse to the surviving universalistic language of the Church.

Notes

1 Examples of this new emphasis might include François Furet, *Penser la révolution française* (Paris, 1978), translated as *Interpreting the French Revolution* (Cambridge, 1981): all citations will be made from this edition; Norman Hampson, *The French Revolution and Democracy* (Reading, 1983); Roger Barny, 'Les mots et les choses chez les hommes de la révolution française', *La Pensée* (1978), 96–115; Marc-Elie Blanchard, *St-Just et Cie – La Révolution et les mots* (Paris, 1980), reviewed by Serena Tuci-Torgussen, *Annales historiques de la Révolution française* 53 (1981), 332–4; Albert Soboul, 'Equality: on the power and danger of words', *Proceedings of the Consortium on Revolutionary Europe* 1 (1974), 13–21; Régine Robin, *Histoire et linguistique*

(Paris, 1973). Also useful as a compendium of examples of the discourse is the older study by François Aulard, *L'éloquence parlementaire pendant la révolution française,* 3 vols. (Paris, 1886).

2 Furet, *Interpreting the French Revolution,* pp. 48–50.

3 E.g. as Robespierre wrote in his paper, *Le patriote français,* 2 September 1792, 'L'âme de la république, c'est la vertu, c'est-à-dire, l'amour de la patrie', quoted in Hampson, *French Revolution and Democracy,* p. 14; Furet, *French Revolution,* p. 59; Barny, 'Les mots et les choses', p. 99.

4 J. G. A. Pocock, 'The concept of language and the *métier d'historien*', E.U.I. Colloquium papers: *The Creation and Diffusion of political languages in early modern Europe,* Florence, 28–30 September 1983, p. 20. We follow here the distinction elaborated by Pocock between *langue* and *parole,* 'language' and 'discourse', the latter being 'a distinguishable language game' with 'its own vocabulary, rules, preconditions and implications' (pp. 3–4).

5 Barny, 'Les mots et les choses', pp. 109–110. The incapacity of the discourse of the Revolution to provide *legitimate* ways of discussing the interplay of sectional interests is discussed at length in Patrice Higonnet, *Class, Ideology and the Rights of Nobles during the French Revolution* (Oxford, 1981), *passim.*

6 The Revolution's use of the Brutus myth is discussed in R. L. Herbert, *David, Voltaire, 'Brutus' and the French Revolution: An Essay in Art and Politics* (London, 1972).

7 Furet, *Interpreting the French Revolution,* pp. 52–3, 128.

8 Examples of that tendency could include Mary Durham, 'Citizenesses of the Year II of the French Revolution', *Proceedings of the Consortium for the History of Revolutionary Europe* 1 (1972–4), 87–109; Darline Gay Levy, Harriet Branson Applewhite and Mary Durham Johnson, *Women in Revolutionary Paris 1789–1795: Selected Documents Translated with Notes and Commentary* (Urbana, 1979); Olwyn Hufton, 'Women in Revolution, 1789–1796', in *French Society and the Revolution,* ed. Douglas Johnson (Cambridge, 1976), pp. 148–64. Non-feminist historians, and non-female ones, do no better: e.g. Albert Soboul, 'Sur l'activité militante des femmes dans les sections parisiennes en l'an II', *Bulletin d'histoire économique et sociale de la révolution française* 11 (1979), 15–26; L. Devance, 'Le féminisme pendant la révolution française', *Annales historiques de la révolution française* 49 (1977), 341–76. The older study by Baron Marc de Villiers, *Histoire des clubs des femmes et des légions d'amazones, 1793–1848–1871* (Paris, 1910), however, contains some perceptive remarks scattered *passim.*

9 Régine Robin, *Histoire et linguistique* (note 1).

10 Pocock, 'Concept of language', pp. 14–17; Furet, *French Revolution,* p. 50; Paul Fussell, *The Great War and Modern Memory* (New York and London, 1975).

11 John Gumperz, *Language and Social Identity* (New York and London, 1983); Robin Lakoff, *Language and Women's Place* (New York, 1975); Barrie Thorne and Nancy Henley, *Language and Sex: Difference and Dominance* (New York, 1975); Leonore Loeb Alder, Judith Orasanu and Miriam K. Slater, eds., *Language, Sex and Gender* (New York, 1979); *Women and Language in Literature and Society,* ed. Sally McConnell-Ginet, Ruth Borkes and Nelly Furman (New York and London, 1980); Howard Giles, W. Peter Robinson and Philip M. Smith, *Language: Social–Psychological Perspectives* (New York and London, 1980); Barrie Thorne, Cheris Kramarae and Nancy Henley, *Language, Gender and Society* (New York, 1983); Cynthia L. Berryman and Virginia Eman, eds., *Communication, Language and Sex* (New York, 1980).

12 N. Hampson, *French Revolution and Democracy,* pp. 14–16 and *passim.*

13 I would like to thank Judith Sklar for her helpful discussion of this point. On Marie Antoinette, see Devance, 'Le féminisme' (note 8); the text of her interrogation by the Revolutionary Tribunal is given in H. Wallon, *Histoire du Tribunal Révolutionnaire de Paris,* 6 vols. (Paris, 1880–2), vol. 1, pp. 296–350.

14 Furet, *French Revolution,* p. 66.

15 Quoted in Applewhite, Levy *et al., Women in Revolutionary Paris* (note 8), p. 93, from Olympe de Gouges, *Les droits de la femme* (Paris, 1791).

16 Marc de Villiers, *Clubs des femmes,* p. 226, shows this line of attack on the 'club des citoyennes révolutionnaires' in 1794.

17 Mme Roland, *Mémoires,* ed. Claude Perroud (Paris, 1905), pp. 63–4, 185, 201, 295–6.

18 Camille Bloch, 'Les femmes d'Orléans pendant la Révolution', *La révolution française* 43 (1902), 46–67, pp. 61–2.

19 Mme Jullien, *Journal d'une bourgeoise pendant la Révolution, 1791–1793,* ed. Edouard Lockroy (Paris, 1881), p. 31 (August 1791). For more discussion of Mme Jullien, see D. Blottière, 'Robespierre apprecié par une contemporaine: Mme Jullien de la Drôme', *Annales révolutionnaires* 4 (1911), 93–6. For Mme Roland, Claud Perroud, ed., *Lettres de Mme Roland,* 2 vols. (Paris, 1900–1), vol. 1, p. 107, and 'Je n'ai plus rien à envier aux antiques républiques' (letter of 5 January 1791); for Mme Cavaignac, *Les mémoires d'une inconnue (1780–1816)* (Paris, 1894), pp. 26–9. As much as Brutus was ritually invoked as the embodiment of republican virtue, so was Cataline invoked as the personification of the anti-republican plot.

20 This dichotomy is also embodied in the art of the period; David's *Brutus* (Louvre), for example, rigidly separates on the canvas the tensely upright figure of Brutus from the swirling, fluid group of his wife and daughters bemoaning the execution of their son and brothers.

21 Furet, *French Revolution,* pp. 56–7.

22 For Hérault de Séchelles' pilgrimage to Mlle Clairon, see his (posthumous) account in the *Magasin Encyclopédique* I (1795), 396–416.

23 *Mémoires d'une inconnue,* p. 29: 'le naturel et le simplicité de ses manières [a] . . . empêche tout rapprochement entre elle et la matronne romaine'.

24 J. P. Faye, 'Langues totalitaires', *Cahiers internationaux de sociologie,* 1964, pp. 36–41.

25 Michael Walzer, *Regicide and Revolution: Speeches at the Trial of Louis XVI* (Cambridge, 1974), pp. 13–14.

26 Roger Dupuy, 'Les femmes et la contre-révolution dans l'Ouest', *Bulletin d'histoire économique et sociale de la Révolution française* 11 (1979), 48–69.

27 Bonnie G. Smith, *Ladies of the Leisure Class* (Princeton, 1981).

28 Examples quoted in Henri Wallon, *Histoire du tribunal révolutionnaire de Paris,* vol. 3, pp. 382, 385, *floréal* 1793.

29 A. Mathiez, 'Catherine Théot et le mysticisme chrétien révolutionnaire', *La révolution française* 40 (1901), 481–518.

30 A. Soboul, 'L'activité militante', p. 26.

31 Lakoff, *Language,* p. 6.

32 Higonnet, *Nobles,* and Barny, 'Mots et choses, p. 114, from D'Oliviers's *Premier suite du vœu national* (Paris, 1790); Soboul, 'Equality', p. 17.

33 See the perceptive comment by S. Alexander, 'Women, Class and Social Difference in the 1830s and 1840's, *History Workshop* 17 (1984), 125–49, especially pp. 134–5, on the relationship between male-led political Revolution, involving the strengthening of

male ego as the Revolution succeeds, and the 'natural' converse process of the confinement of the female ego.

34 Levy, Applewhite and Johnson, *Women,* p. 93, from Olympe de Gouges, *Les droits de la femme* (Paris, 1791); Wallon, *Tribunal,* vol. 2, p. 245.
35 Bonnie G. Smith, *Ladies of the Leisure Class, passim* and conclusion.

7

LIFE-PATHS

Autobiography, science and
the French Revolution

From: Michael Shortland and Richard Yeo (eds.), *Telling Lives in Science: Essays on Scientific Biography* (Cambridge and New York: Cambridge University Press, 1996), 85–102. © Cambridge University Press. Reprinted with permission.

Introduction

François-René de Chateaubriand, the great French romantic, spoke for an entire generation when in his autobiography he described the French Revolution as a 'river of blood which separates for ever the old world in which you were born from the new world on whose frontiers you will die'.[1] This sense of the French Revolution as a dark and threatening chasm in time, so different from the joyful *novus ordo saeculorum* of the triumphant New World revolution of 1776, was undoubtedly one of the major reasons for the outpouring of autobiography which is characteristic of that era.[2] This was so not only because autobiography seemed the ideal vehicle, in many cases, to vindicate the actions of the troubled revolutionary era, but also because autobiography became a way of asserting the continuities of existence and identity across the rupture between the pre- and post-revolutionary worlds. Because of this, it is difficult, and misleading, to see the autobiographies of this time as 'private' documents: rather, they are involved with the reshaping of the meaning of the revolutionary era, precisely because they are the medium through which many in France worked through the specific nature of their investment in, and exploration of, the 'new world' of post-revolutionary time.

Possibly nowhere did autobiography fill this need more urgently than in the case of the scientific élite. As the old élites emigrated, were executed, or were forced out of employment, the élite of science became one of the few sources of

---

1 Chateaubriand (1958), 165.
2 Dekker (1969), 61–72; 63–4; Tulard (1971) lists 794 items of printed autobiography. The collection omits MS autobiography, and has little interest in the scientific community. The growth of introspective private writing of all kinds is described in Bryant (1978).

the technical expertise so greatly needed by successive French regimes forced to mobilise the resources of France in the warfare which began in 1792 and was not to end until 1815. It was this which allowed the hitherto unprecedented rise of scientific men to positions of real political and administrative power.[3] This capture of power outside the narrow area of technical expertise was aided by the fact that the public culture of the Revolution legitimated its break with the past by appeals to values which science was uniquely fitted to endorse: by appeals to 'Nature', and to 'Reason', as distinct from the key cultural values of the former monarchy, of historicity, heterogeneity and artificiality. For a revolution trying to find ways in which it could do that which the Jacobin regicide Louis-Antoine St Just defined as impossible for monarchy and 'reign innocently', science was the ideal cultural legitimation, and its practitioners therefore legitimate bearers of public authority.[4]

Science, and its individual practitioners, thus underwent a dramatic change in cultural field and force. At the same time many individuals within the scientific community experienced extreme reversals of personal fortune. It is thus hardly surprising that autobiographical writing should flourish among the French scientific élite of the Revolutionary period, as a means of coming to terms with the unbridgeable spaces opening up between past and present lives and roles. This outburst of autobiographical writing was also stimulated by a specific institutional pressure. Georges Cuvier and Jean-Baptiste Delambre, the Permanent Secretaries of the First Class of the Institut, made it a practice to request autobiographical material from those whose funeral orations or *éloges*, they might later have to write.[5] However, the fact that many, though not all, autobiographies were produced on request from those whose task it was to manufacture the 'official history' of science, does not seem to have made such autobiographies bland or impersonal; far from it. In fact, one of the most interesting features of the autobiographies is their very difference from the official lives of their subjects. This is a topic to which I will return.

But the fact that autobiography became in any way a resource for *éloges*, in itself signals the increasing incidence and acceptability of autobiographical writing among the scientific élite. For the efflorescence of autobiography amongst this group was not unchallenged and did not have deep historical roots. More traditional thinkers still insisted that for men of learning, autobiography was not only unnecessary, but was even a logical contradiction to the ideal of the identity of life and work. The absolute identity of life and work, and hence the inutility of autobiography, for all *savants*, was still being maintained as late as 1807, when Pierre-Charles Levesque maintained of the classical scholar Apollonius of Tyanus that 'it

3 Outram (1984 and 1980).

4 For artifice in the public life of the old Regime, see Sennett (1974), 1–99; Louis-Antoine St Just, speech of 13 November 1792, quoted in Walzer (1974), 124, from *Archives Parlementaires,* première série, 53 (1792), 391.

5 Autobiographies produced in this context are listed in Outram (1978).

seems that following the precept of an ancient philosopher he wanted to hide his life; but he completely revealed it when he published his work'. This identity could only be damaged by the process of introspection so necessary to autobiography, with its implication that the path from past to present, from the life to the work, was likely to be broken, uneven, and full of the unforeseen.[6] The social and intellectual conditions of autobiography in the French scientific élite could not therefore rely on a settled social response. The age itself challenged the idea of the centrality of the individual; the professional ethos was unsure of its reception of autobiography.

The French Revolution had produced perceptions of the fundamental category of autobiography, time itself, which were new and disturbing. The idea of the Revolution as a rupture in time, was accompanied by an unparalleled sensation of time as having speeded up.[7] Perhaps we are seeing here the first symptoms of what modern theorists of the information revolution have described as the collapse of the categories of real time. In this situation, writing autobiography, which depends on the unfolding of lived time, represented severe challenges. This paper will examine several examples of autobiography produced by members of the French scientific élite to show how autobiographical writing responded to these varied pressures. The examples discussed in this paper are drawn from only a small fraction of extant autobiographical writings. These are listed at the end of this chapter; however many more undoubtedly await discovery in family and public archives. Most of these autobiographies were produced by members of the French scientific élite, members of the Academy of Sciences, or the National Institute. Far from being private documents, these autobiographies were often produced to respond to public or institutional demand. This is an important point to make because autobiography can often be used as the primary source for biography. It is thus important to realise that autobiographies are not raw data but reflect the traces of the many pressures on autobiographical writing; as a genre, they are a precious resource but also tell us much about the problems of the scientific enterprise in revolutionary and post-revolutionary France.

Journeying, knowledge and self-knowledge

To confront the central metaphor of my title: the image of life as a path to be followed is a very ancient one, which was transformed in the Christian West into the image of man as a lifelong pilgrim. Debate raged by the end of the Middle Ages

---

6 Even the 'autobiography' of G. B. Vico maintains an identity between 'life' and 'work', in spite of the fact that it was also Vico who began to generate and collect scientific autobiography (Fisch and Bergin (1944)); Levesque (1807), xviii: 'il semble que, soumis au précepte d'un ancien philosophe, il ait voulu cacher sa vie; mais il l'a dévoilée toute entière en publiant ses ouvrages'.

7 E.g. Mme Roland exclaimed, 'Here we live ten years in twenty-four hours': Perroud (1900–1901); Ramond remarked in the National Assembly 'the last six months of 1789 were worth an ordinary year' (2 January 1792) quoted in Aulard (1885), I, 89.

over the proper conduct of the pilgrim, and involved issues of particular importance for the sciences. At the heart of the debate on pilgrimage was the problem
of curiosity.[8] Pilgrims both actual and metaphorical were repeatedly cautioned
not to let their eyes stray from their paths, if they were to arrive at their goal.
This is an image still active in Protestant texts of the seventeenth century such
as John Bunyan's *Pilgrim's Progress*. By the eighteenth century, this debate had
been secularised and inverted. Journeying, and the growth of knowledge in and
about life, were repeatedly connected. Jean-Jacques Rousseau's *Emile* (1762), for
example, produced a strong defence of straying from the beaten path as the best
way to acquire knowledge:

> The only way I can imagine travelling more pleasant [than horseback]
> is to go on foot. . . . You can see all the countryside, you can turn left or
> right, you can have a proper look at anything which interests you, and
> you can stop wherever there is a view. If I notice a river I walk along its
> banks; if I see a deep wood, I enter its shade; a cavern, I can enter it; a
> quarry, I can look at the rocks. To travel on foot is to travel like Thales,
> Plato, and Pythagorus. I really can't understand how an enquiring person
> could decide to travel in any other way.[9]

While this passage makes very clear the eighteenth-century links between straying from a set path and observational curiosity, later, the very title of Rousseau's
Reveries of a Solitary Walker[10] (1782) was to make the connection between cover-
ing the ground and the work of introspection even more complete. It was a metaphor which also allowed the linkage of the life and work to go on being made at
another level not by cutting out the life, but by seeing it as a web of movement,
curiosity and introspection, which came together in a scientific vocation.

It was in this way that movement and autobiography became inextricably
linked, and made their way into the autobiographies of men of science. Sometimes, accounts of journeying are also ways of dealing with the fault lines which
described an entire life. For Cuvier, for example, the inner movement from childhood to adolescence was also a movement from one language to another, all
encapsulated in an actual journey from his birthplace in provincial Montbéliard,

8 Ladner (1967); Zacher (1976).
9 Rousseau (1964), 522–3: 'Je ne conçois qu'une manière de voyager plus agréable que d'aller à
cheval; c'est d'aller à pied. On observe tout le pays; on se detourne à droit, à gauche; on examine
tout ce qui nous flatte; on s'arrête à tous les points de vue. Aperçois-je une rivière, je la cotoie; un
bois touffu, je vais sous son ombre; une grotte, je la visite; une carrière, j'examine les mineraux
. . . Voyager à pied, c'est voyager comme Thalès, Platon et Pythagore. J'ai peine à comprendre
comment un philosophe peut se resoudre à voyager autrement'. A further panygyric to curiosity,
linking it to movement, is on p. 185.
10 Written between 1776 and 1778, and published in 1782. The same connection between walking,
'pilgrimage' and self-knowledge persisted into the following century: Thoreau (1862).

to school in cosmopolitan Stuttgart. Such was the impact of this linked inner and outer journey, that even forty years later, Cuvier was unable to recall it without experiencing

> a sort of terror at this journey, which I made in a little carriage squashed between the duke's chamberlain and his secretary, who I was really in the way of, because there was barely room enough for them as it was. For the whole journey they only conversed in German, which I didn't understand a word of, and hardly spoke a single word of encouragement or consolation to me.[11]

In the case of Bernard Germaine Lacépède, journeys too are paths into different self-representations. As his autobiography details his journeys from Paris to Agen, from Agen to Paris, from Versailles to Germany, it also details his travelling between the different roles which one after the other filled his life before his vocation as a man of science became firm. Local notable in Agen and erudite, in Paris, musician, in Germany, soldier and diplomat, Lacépède's vocation only became clear when he reached the end of a final voyage, from Paris to his wife's property at Montlhéry:

> It was in the beautiful countryside which surrounds this village, in the charming meadows, in the little woods near the old country house, that I thought through the plan for my natural history, and here too that I wrote most of its first volume.[12]

The metaphor and reality of travel as being and becoming, of life as a curious exploration of many paths, is extended in the autobiography of Ramond de Charbonnières.[13] A tumultuous early life, always following 'un autre horizon',[14] was almost brought to a close in the prisons of the Terror, where Ramond compared himself to Robinson Crusoe, the voyager and spiritual pilgrim, who in losing all upon his desert island, finds the spiritual insight he had previously disdained.[15]

---

11 The autobiography of Cuvier is MS Flourens 2598 of the Library of the Institut de France: 'une sorte d'effroi, à ce voyage que je fis dans une petite voiture, entre le chambellan et le secrétaire du Duc, que je gênais beaucoup, parcequ'il y avoit à peine de la place pour eux, et pendant toute la route ne se parlèrent qu'en allemand, dont je n'entendais pas un mot, et m'addressèrent à peine deux paroles d'encouragement et de consolation'.

12 Hahn (1975), 61, 64: 'C'était dans les belles campagnes qui environnent ce village, dans ses charmantes prairies, dans les bosquets de l'ancien château, que j'avais médité le plan de mon *Histoire naturelle,* et composé une grande partie du premier volume de cette Histoire'.

13 Dehérain (1905).

14 Ibid., 124.

15 Ibid., 128. 'ah! le beau livre que je vous ferais, nouveau Robinson que j'ai été . . . '.

In travelling towards their making and remaking of self, autobiography writers of the eighteenth century were not alone. They travelled with a whole range of different reference figures on which to base the struggle for identity. The eighteenth century was an age of mimesis – an age when modelling the self on some figure of 'representative virtue' was a commonplace psychological exercise.[16] For no group was it an exercise more important than for the scientific community, the one most in need of models to negotiate the transition between old and new worlds. This means also that in considering eighteenth-century autobiography, we are often not merely considering the life of the ostensible subject, but also the relationship between that life and the role models which he or she adopts.

But the search for role model also transmitted conflict into the autobiographical form. Role models could be the stern figures of classical antiquity, or the newer heroes of science and virtue such as Newton or Benjamin Franklin. They could also be taken from the emotional narratives which were eighteenth-century best-sellers, novels like Richardson's *Clarissa* (1747–8) or Rousseau's *Julie ou la Nouvelle Héloïse* (1761). These were models whose actions put them fairly at odds with the restraint and disinterest expected of the ideal man of science; they also encapsulated, in their exaltation of emotion, a whole response to the natural world which would be characteristic of Romanticism. This is a conflict epitomised by the difference between the Rousseau of the *Confessions*, of *Julie* and of the *Reveries*, and the Rousseau of *Emile* and the *Social Contract*.[17]

These were conflicts of which autobiography writers were well aware. Ramond, for example, consciously modelled his autobiography in novel form to avoid both the very question of the identity of life and work, as well as to avoid the political commitments implicit in the sterner genre of the *Mémoire* or 'History of my own times'. He wrote to his friend St Amans, who had urged him to produce an autobiography:

> you want me to stop writing about Nature, and become part of the throng of those who are constantly retailing the slightest doings of our contemporaries. You need a 'life and times', you aren't put off by all this; but my life and times? Certainly they would amuse a lot of people, even though it would be my puny self who would be playing the hero for, as well as the great and small events which I would be describing as so many others have done, the ups and downs of my own personal fortunes would read like a novel and have all the pulling power of the genre.[18]

---

16 Tatin (1985).

17 Outram (1989).

18 Dehérain (1905), 122: 'vous voudriez que j'abandonasse cette partie des ouvres de la nature qui va sans dire, pour me mêler dans la foule de ceux qui bavardent dans tous les sens sur les faits et gestes des hommes de leur tems. Il vous faut des mémoires: vous n'êtes pas dégouté: mes mémoires! certes ils amuseraient beaucoup de monde, quand même mon chétif individu y serait sur le premier plan, car au milieu des grands et petits evenemens dont j'aurais à essayer l'esquisse après tant d'autres, les vicissitudes de ma propre fortune constitueraient à elles seules une manière de roman qui aurait tout l'intérêt du genre'; Trahard (1936), 31–6.

The autobiography which resulted is, in fact, written as a picaresque novel. To take another example, Lacépède wrote some passages of his autobiography in a way which recalls nothing more than the reactions to scenery of the characters in *Julie*:

> Seated on the ruins which surround the high tower at Montlhéry, looking out over an immense prospect, seeing the distant outline of the monuments of the capital; or lying on the flowered grass, in the shade of whispering poplars on the shores of the great lake at Marcoussis, or strolling under the green vault of the vast and lonely forest on the mountain tops around the lake, I love to meditate on the wonderful effects of the power of Nature . . . Given over to these sublime thoughts, dragged away by great conceptions, seduced by these magical scenes, I forgot the world, and saw nothing but the universe.[19]

Here, Lacépède portrays his relation with Nature as very different from that of the self-sufficient, austere man of science beloved of the Stoic tradition. His attitude is defined by emotion, not reason, by a blurring of the boundaries between himself and Nature due to an emotional connection with landscape. In this part of his autobiography, Lacépède becomes the man of sentiment so beloved of the eighteenth-century novel. The smooth self-possession of the Stoic ideal vied for place with Rousseau's ideal of autobiography as a total revelation, and life as the fashioning of emotion.[20] Lacépède in fact uses autobiography as a story of the clarification of commitment, to science, a commitment which seems, at many points in the autobiography, to be nearly overwhelmed by the pressure of unresolved emotion.

Autobiography: vehicle for commitment

Clearly, autobiography was of great utility in clarifying the commitments of the writer. And it is not surprising that this psychic function should be mirrored by a focus on episodes of commitment within the autobiographies themselves. To some extent, such episodes can be recounted as moments of absorption, like the one just quoted from the autobiography of Lacépède; or the one contained in the otherwise very different autobiography of Georges Cuvier, who relates how

---

19 Hahn (1975), 69; 'Assis sur les ruines qui environnent la haute tour de Montlhéry, dominant sur un pays immense, découvrant de loin le faîte des superbes monuments de la capitale, ou couché sur un gazon fleuri à l'ombre de peupliers inspirateurs et sur les bords du grand étang de Marcoussis, ou me promenant sous les vôutes de verdure formés par les vastes et solitaires forêts qui couronnaient les montagnes autour de cet étang, j'aimais à méditer sur les admirables effets de la puissance de la nature . . . Livré à des conceptions élèves, entraînée par ces grandes pensées, séduit par ces tableaux magiques, j'oubliai le monde, je ne voyais plus que l'univers.'

20 For the influence of Rousseau's idea of autobiography as complete self-revelation, see Outram, *The Body* (note 17), 149–51.

Buffon's *Natural History* completely absorbed him as a child: 'All the pleasure I had as a child was in copying its pictures and colouring them according to the captions'.[21] Such moments of intense focus, with their links to later self-dedication, have obvious religious echoes.[22] And it is worth mentioning at this point how closely autobiography had previously been linked with the history of individual spirituality. Indeed, in many national traditions, autobiography *was* the recounting of spiritual pilgrimage and conversion.[23] In spite of the secularisation of many features of autobiography during the eighteenth century, and its mixing, as we have seen, with other forms of narrative of individual lives it remained marked by the earlier history of the genre. Moments of epiphany, of absorption in Nature, in scientific autobiography have the same role as conversion moments in spiritual autobiography: they resolve the antagonism of the self and the world. What brings optimism to Lacépède's account is the way in which his autobiography is structured as a journey *towards* such an epiphany: namely, his absorption in Nature. What gives Cuvier's its pessimism is the movement of his account away from 'absorption' in infancy, into a relentless confrontation between himself and the 'world', power and patronage, conceived as something whose control and manipulation must be so endlessly recreated that such contact with Nature, such blurring of the lines of the self, can never recur. The precision with which Cuvier tells us the details of this conflict is precisely what makes Cuvier's autobiography such an important source in the history of scientific politics and patronage.

Nor is this the only way in which even the post-Rousseau autobiography retains traces of the religious associations of the genre. Linking science, religion and autobiography was the notion of vocation. Often, the finding of a vocation is linked with the rejection or loss of a biological father, and the finding of new, 'social' father, the patron, who guides the novice man of science and acts as his role model. Just as in early Christianity, the finding of a religious vocation, the rejection of the biological family, and the adoption of a saint as patron, were often inextricably mixed events.[24] This theme is particularly marked in Lacépède's

---

21 Discussed in Outram (1984), 19, and quoted from MS Flourens 2598 of the Institut de France: 'Tout mon plaisir d'enfant était d'en copier les figures, et de les enluminer d'après les descriptions'. For the importance of the portrayal of such moments of absolute concentration of the forces of the personality, see Fried (1980).

22 For modern equivalent experiences, see Terres (1961); Cobb (1977); Austin (1931), 24–5: 'It was a summer morning, and the child I was had walked down through the orchard, and come out on the brow of a sloping hill where there were grass and a wind blowing, and one tall tree reaching into infinite immensities of blueness. Quite suddenly after a moment of quietness there, earth and sky and windblown grass, and the child in the midst of them came alive together with a pulsing light of consciousness. There was a wild foxglove at the child's feet, and a bee dozing about it, and to this day (1931) I can recall the swift inclusive awareness of each for the whole – I in them and they in me and all of us enclosed in a warm lucent bubble of livingness'.

23 Dekker (1969); Niggl (1977); Roodenburg (1985).

24 Theis (1976); Silverman (1974).

autobiography, which describes many moments of 'adoption' by individual patrons. Lacépède begins with his childhood memories of the bishop of Agen, who treated him 'as if I had been his child',[25] as a young man Buffon regarded him 'comme son fils', and after the devastating event of his own father's death in 1783,[26] he immediately formed a close relationship with the Gauthier family, who virtually adopted him.[27] Later, he was to marry the widowed Mme Gauthier, and it was, as we have seen, on her estate, that his vocation as a natural historian was solidified: absorption in Nature, and reintegration into a second, 'social' family, came together. A similar process was undergone by the young chemist Eleuthère Dupont de Nemours.[28] So common was it, that Cuvier's autobiography is at great pains to explain why it was that, unlike his peers, he was unable to record the same moments of adoption and confirmation by a patron. Clearly, such moments of rupture with natural family, and readoption by 'social' family or patron taking the role of father fitted easily with Stoic ideals of the essential disinterest of the man of science, and with a vision of his role as outside social production.

Autobiography under the conditions of revolution

But these were roles maintained against great odds. The period of the Revolution saw a collapse, for all sections of the French élite, of the patronage networks stemming from members of the former privileged orders. With these networks in disarray, the search for patronage and recognition became ever more uncertain, and this uncertainty was compounded by the fact that the role of the man of science in public life was in rapid transition. It was also a victim of the insecurity of Revolutionary politics, and the collapse of all accepted forms of status and social recognition. For those who fell out of political favour, the problem of maintaining a consistent face to the world thus became almost insuperable. Because of this, autobiographies from this period often detail dramatic shifts in social status, shifts which are perceived as changes in dramatic role. As Ramond comments, 'I could see the storm clouds coming nearer. I hoped to be able to be just a spectator; but events quickly made me an actor in the drama'.[29] For many in the scientific community, as outside it, survival itself often depended on the rapid conversion from

25 Hahn (1975), 56.
26 Ibid., 62: 'La douleur que j'eprouvai fut affreuse . . . et ma vive et bien juste affliction me donna une maladie nerveuse qui ne me permit de quitter Agen pour revenir à Paris, que vers le milieu de l'été 1784'.
27 Ibid., 63: 'Ils me permirent de les regarder comme un frère et comme une soeur, et après les pertes irréperables que j'avois faites, combien ne dus-je pas me féliciter de retrouver en eux une seconde famille!'
28 Dupont de Nemours (1906): 'Je n'étais qu'un enfant lorsqu'il me tendait les bras. C'est lui qui m'a fait un homme'. Cuvier's autobiography is virtually unique in its absence of such moments.
29 Dehérain (1905), 126: 'je voyais approcher l'orage. Le rôle de spectateur était celui où j'espérais pouvoir me renfermer: la force des choses me mit bientôt au nombre des acteurs'.

the roles of the *Ancien Régime* into those less conspicuous, or more politically correct.[30] Very often, such role changes were associated with a fall in social status. The naturalist Louis-Augustin Bosc, for example, too closely associated with the Girondins in 1793, recounts in his autobiography how he fled Paris, and adopted the clothes and life-style of a working man.[31]

Life under the revolution placed a survival value on mimesis. But these episodes of role-shift also interrupt the flow of many autobiographies from past to present, doubt to resolution. For some, the interruption in their role as scientific men, and accompanying status, was too great to bear.[32] For others, such as Ramond, science emerged as the lifeline which held identity together.[33] But the enforced roles of the Revolution inject an incongruous note into the strands which we have hitherto identified as part of the heritage of scientific biography. The idea of life as a journey, the idea of vocation as achieved in moments of 'epiphany', of supreme absorption, and as realised by the movement between natural family and social family, were all perturbed by the incongruous roles forced upon many men of science by the Revolution. Under such roles, the speaking subject of the autobiography becomes literally masked, the path of life diverted. The chasm in time which was for many the Revolutionary experience, also becomes a chasm in autobiographical narrative.

In registering this disruption, the autobiographical form was also marking another major change in life-paths. Before the revolution, the possession by a single individual of multiple social roles was not viewed as exceptional. This is precisely why many autobiographies detail a twisted and divergent path towards their subject's vocation. Nor does it surprise us that men like the chemist Antoine-Laurent Lavoisier should hold multiple social roles, roles which would now be seen as grossly incompatible should any individual try to combine them in the twentieth century. Landowner, financier, local administrator and research chemist, were combined in Lavoisier's person, and his case is far from unique.[34] The teasing out and separation of roles from each other was to be a characteristic of the nineteenth century, one which was to help to make autobiography a different exercise, partly by changing the social conditions in which autobiography was written and received.

A final point to note about these autobiographies is that of their gendered Nature. All extant scientific autobiographies of this period known to me are written by

---

30 Outram (1983).
31 Bosc's autobiography is cited in an edited version in Perroud (1905): 'je m'habillai à la sans-culotte, travaillai à la terre, au bois . . . fis moi-même ma cuisine'.
32 See autobiographical writings of the astronomer Cassini, discussed in Outram (1983), 263–44, using autobiographical material quoted in Devic (1851), 290–91.
33 Outram (1983), 256.
34 Little attention has been paid to this problem: see however, Vergnaud (1945–1955); Outram (1984), 93–117.

men.[35] They recount experiences which were only available to men in this period, experiences such as the search for membership of élite scientific institutions, or the holding of public office. But in another way, even the autobiographies of men of science manifest much of the confusion and contradiction entering into pre-revolutionary experiences and discourses of gender. Writings such as Lacépède's, for example, are suffused with the emotionality of the novel, which, as literary critics have noted, increasingly made the literary field the producer of common emotional patterns and role models between the genders.[36] On the other hand, many other autobiographies, still revolve around the very masculine stereotype of the Stoic man of science. This is key to Ramond's self-portrayal at the end of his captivity, for example, as a 'new Robinson', denuded of the world's goods, and with the wish, after the passions of the Revolution, to focus only on the objectivity of Nature.[37] The Revolution itself was to see a dramatic hardening of the division between the genders, a hardening whose consequences for future politics, and for the way in which science was professionalised, have only recently come to be seriously appreciated.

It is also the case that these male autobiographies tend to produce a different account of the 'turning-points' of life. In many autobiographies the stage of life that makes the transformation from child to adult, is marked, as we have seen, by a moment of 'adoption' by an older, usually male patron, as well as moments of supreme absorption in Nature itself. The definition of vocation in other words, in most male autobiographies, is defined as finding a new 'father', becoming the 'child' again. Little stress is placed in many of the autobiographies on marriage or paternity, usually mentioned only in passing, if at all. In this respect, the autobiographies are very similar to the official contemporary biographies or *éloges*. This is a very different world from the life cycle events such as puberty, marriage and maternity, which we find dominant in contemporary female autobiography, such as that produced by Mme Roland.[38] These are differences which reflect the very different social insertion and self-images of men and women at this time. Research into actual life situations amongst the scientific élites of this period reveals the very ambiguous role of marriage for the man of science, and the heterogeneous character of the marriage partners chosen.[39] Marriage in other words, was a social practice whose integration with the self-image of the man of science, was very problematic. Very often, this exclusion of the marriage relationship, or other love relationships, from both biography and autobiography, results in what would now

---

35 Women, however, often played a significant role in the editing of manuscript autobiographical material: the serious excisions in Cuvier's autobiography perpetrated by Mme Cuvier after his death in 1832, are a case in point. The influence of women on the preparation of documentary evidence was all the greater in the period of the Revolution due to the high number of male deaths.

36 Eagleton (1982); Outram (1989), 138–41. For contrary argument see Nussbaum (1989).

37 Dehérain (1905), 128.

38 Perroud (1905); Geiger (1986).

39 Outram (1987), 19–30.

be seen as bizarre displacements of personal emotion into the fabric of scientific writing. This means that it is difficult both to talk of 'scientific' writing in the same way that we would do now, as descriptions of investigations into Nature devoid of personal emotional content; and, that it is also wrong to draw rigid lines between autobiography and other forms of writing. Just as scientific professional roles were only starting to disentangle themselves from the multiple roles taken for granted by élites, so it was that autobiography itself was yet to separate out from other forms of scientific prose. Lacépède, for example, remarks matter-of-factly in his autobiography about his grief at the death of his wife in 1801: 'There is an expression of my grief in the fifth volume of my *Natural History of Fish*'.[40] It is clear from these examples that the impermeable divisions between the work and the life, Nature and its observer, objectivity and subjectivity, the product of 'male science' according to much feminist writing, were not yet in place in the writings of these male scientists.[41]

There is a real contradiction here between this play of violent personal emotion into scientific writing, and the ideal of the man of science as a calm and disinterested figure, moved only by the encounter with Nature.[42] This contradiction is a genuine one and can be traced back to the porosity of the autobiographical form itself, which was here only reflecting the cacophony of different models of identity with which Enlightenment men and women were surrounded, and tried to internalise. In turn, this reflected increasing problems in this period in defining both public and private roles for individuals. France at the end of the eighteenth century saw increasing competition to define the public realm, and thereby define who the actors in that public realm were to be:[43] who, in other words, were to replace the monarch as sole public actor. Increasingly, the definition of sovereignty, of public identity, became conflated with that of self-possession, self-definition. This may also be one of the reasons why most scientific autobiographies of this period are written by men who were far from the end of their life-span. These autobiographies were not primarily written to display the distillation of a lifetime search for achievement and self-knowledge; they were typically written in mid-life, to *create* public personalities and manifest self-possession. They were documents written to induce change, not to record passively that which had already occurred. In this way, they contrast strikingly with autobiographies by English men of science such as Humphry Davy, almost all of which were produced later in life.

---

40 Hahn (1975), 74; Lacépède adds to his autobiography an extract from his *Histoire de l'Europe* (1816) in which he recalls his grief on the death of his daughter-in-law (ibid., 83). Among many other possible examples could be cited Dolomieu (1793), 41, which contains his eyewitness account of the murder of his friend the Duc de la Rochefoucauld.
41 For the social consequences of the absence of these divisions see Outram (1987).
42 Sonntag (1974); Outram (1978).
43 Outram (1989), 68–89.

But in saying all this we are making something more than the banal point that scientific autobiography, like any other form of art, reflects the strains of the age which produced it. We are also pointing to one of the very reasons for the resurgence of the scientific community in social, cultural and political authority in the 'new world' which emerged after 1789. To have resolved all these complex and contradictory elements into the search for a human identity which manifested the contradictory elements of the vocation of science, was to have absorbed too, a responsiveness to change, a capacity to intervene in the chaotic public culture of post-revolutionary France. To be able to offer its practitioners to the public both as practitioners of Stoic calm, and as mobilising an emotional relationship with Nature, was to offer a resource unique in the public culture of the age. They could do this because these autobiographies were still structured around the turning points of vocational choice, rather than around moments of discovery in science which were to become important in nineteenth-century biographies and autobiographies. Whereas eighteenth-century autobiographies are powerful vehicles which talk about mobilising a *life*, the nineteenth-century autobiographies focus on a single 'Eureka' *moment* of discovery, and fail to engage with their surrounding culture as a whole.

Bibliography

Abir-Am, P. and D. Outram (eds) (1987) *Uneasy Careers and Intimate Lives: Women in Science 1789–1979*. New Brunswick: Rutgers University Press.

Aulard, F. (1885) *L'Eloquence Parlementaire pendant la révolution*, 3 vols. Paris: Hachette.

Austin, M. (1931) *Experiences Facing Death*. London: Rider.

Bryant, M. (1978) Revolution and introspection: the appearance of the private diary in France. *European Studies Review*, 8, 259–72.

Chateaubriand, F.-R. de (1958) *Mémoires d'outre-tombe* (ed. Maurice Levaillant and Georges Moulinier), 2 vols. Paris: Pléiade.

Cobb, E. (1977) *The Ecology of Imagination in Childhood*. New York: Columbia University Press.

Dehérain, H. (1905) Une autobiographie du baron Ramond, membre de l'Académie des Sciences. *Journal des Savants*, March, 121–9.

Dekker, R. M. (1969) Ego-documents in the Netherlands, 1500–1814. *Dutch Crossing: A Journal of Low Countries Studies*, 39, 61–72.

Devic, J.-F. S. (1851) *Histoire de la vie et des travaux scientifiques et littéraires de J.D. Cassini, IV*. Clermont.

Dolomieu, D. (1793) Mémoire sur la constitution physique de l'Egypte. *Journal de Physique*, 32, 41–2.

Dupont de Nemours, H. A. (ed.) (1906) *L'enfance et la jeunesse de Dupont de Nemours*. Paris.

Eagleton, T. (1982) *The Rape of Clarissa: Writing, Sexuality, and Class Struggle in Samuel Richardson*. Oxford: Oxford University Press.

Fisch, M. H. and Bergin, T. G. (eds.) (1944) *The Autobiography of Giambattista Vico*. Ithaca: Cornell University Press.

Fried, M. (1980) *Absorption and Theatricality: Painting and Beholder in the Age of Diderot*. Berkeley: University of California Press.

Geiger, S. N. G. (1986) Women's life-histories: method and content. *Signs*, 11, 334–51.

Hahn, R. (1975) L'autobiographie de Lacépède retrouvée. *Dix-huitième siècle*, 7, 49–85.

Ladner, G. B. (1967) *Homo viator*: medieval ideas on alienation and order. *Speculum*, 32, 233–59.

Levesque, P.-C. (1807) *Notice historique sur Le Grand d'Aussy . . . lu à la séance publique du 15 messidor an X*. In Legrand d'Aussy, *Vie d'Apollonius de Tyane*. Paris: Collin.

Niggl, G. (1977) *Geschichte der deutschen Autobiographie im 18. Jahrhundert: Theoretische Grundlegung und literarische Entfaltung*. Stuttgart: Metzler.

Nussbaum, F. A. (1989) *The Autobiographical Subject: Gender and Ideology in Eighteenth Century England*. Baltimore: Johns Hopkins University Press.

Outram, D. (1978) The language of natural power: the funeral *éloges* of Georges Cuvier. *History of Science*, 16, 153–78.

Outram, D. (1980) Politics and vocation: French science, 1793–1830. *British Journal for the History of Science*, 13, 27–43.

Outram, D. (1983) The ordeal of vocation: The Paris Academy of Sciences and the Terror. *History of Science*, 21, 252–73.

Outram, D. (1984) *Georges Cuvier: Science, Vocation and Authority in Post-revolutionary France*. Manchester: Manchester University Press.

Outram, D. (1987) Before objectivity: women, wives and cultural reproduction in early nineteenth-century French science, in Abir-Am and Outram (1987), pp. 17–35.

Outram, D. (1989) *The Body and the French Revolution: Sex, Class, and Political Culture*. New Haven: Yale University Press.

Perroud, C. (ed.) (1900–1901) *Correspondence de Mme Roland*. Paris: Imprimerie National.

Perroud, C. (ed.) (1905) *Mémoires de Mme Roland: Nouvelle Edition Critique*. Paris: Imprimerie National.

Roodenburg, H. W. (1985) The autobiography of Isabella de Moerloose: sex, childbearing, and popular belief in seventeenth-century Holland. *Journal of Social History*, 18, 517–40.

Rousseau, J-J. (1964) [1762] *Emile ou de l'Education* (ed. François and Pierre Richard). Paris: Garnier Frères.

Sennett, R. (1974) *The Fall of Public Man*. Cambridge: Cambridge University Press.

Silverman, S. M. (1974) Parental loss and the scientist. *Science Studies*, 18, 259–64.

Sonntag, O. (1974) 'The motivations of the scientist: the self-image of Albrecht von Haller, *Isis*, 55, 336–51.

Tatin, J. J. (1985) Relation de l'actualité, reflexion politique, et culte des grandes hommes dans les almanachs de 1760 à 1793. *Annales historiques de la Révolution Française*, 57, 3–16.

Terres, J. K. (ed.) (1961) *Discovery: Great Moments in the Lives of Outstanding Naturalists*. Philadelphia: Lippincott.

Theis, L. (1976) Saints sans famille? Quelques remarques sur la famille dans le monde franc à travers les sources hagiographiques. *Revue historique*, 255, 3–20.

Thoreau, H. (1862) Walking. *The Atlantic Monthly*, 9, 659–67.

Trahard, P. (1936) *La sensibilité revolutionnaire*. Paris; Boivin.

Tulard, J. (1971) *Bibliographie critique des mémoires sur le consulat et l'Empire*. Geneva; Droz.

Vergnaud, M. (1945–55) Un savant pendant la révolution. *Cahiers internationaux de sociologie*, 17–18, 123–39.

Walzer, M. (1974) *Regicide and Revolution: Speeches at the Trial of Louis XVI*. Cambridge: Cambridge University Press.

Zacher, R. A. (1976) *Pilgrimage and Curiosity*. Baltimore: Johns Hopkins University Press.

Scientific autobiographies of the French Revolution: a preliminary listing of printed and manuscript materials

Adanson, Michel, 1727–1806
MS. autobiography 17 June 1775, Library of the Institut de France Fonds Cuvier 3185/6.
Histoire naturelle du Sénégal (Paris, 1757), pp. 1–90 relate life to 1754.
Familles des plantes (1763), I, clvii – viii.

Ampère, André-Marie, 1775–1836
Mme C. Cheuvreux (ed.), *André-Marie Ampère: Journal et Correspondance 1793–1804* (Paris, 1869).
Anon., 'L'autobiographie d'Ampère'. *Bulletin de la Société des Amis de André-Marie Ampère*' (1955), 16, 1–27.

Bonnet, Charles, 1720–1793
Raymond Savioz (ed.), *Mémoires autobiographiques de Charles Bonnet de Genève* (Paris, Vrin, 1948).

Bory de St-Vincent, Jean Baptiste, 1780–1828
Mémoires sur les cent-jours pour servir d'introduction aux souvenirs de toute ma vie (Paris, 1938).

Bosc d'Antic, Louis-Augustin Guillaume, 1759–1828
MS. autobiographie, n.d. Library of the Institut de France Fonds Cuvier 3157/1. Copy by Mme Cuvier.
A fragment of MS autobiography in Library of Muséum National d'Histoire Naturelle Paris, probably written before 1815, with a postscript after 1815.
This fragment published in Claude Perroud (ed.), *Mémoires de Mme Roland, nouvelle édition critique* (2 vols, Paris, 1905), II, 450–61.
The Fonds Cuvier version published by Claude Perroud, 'Le roman d'un Girondin'. *Revue du dix-huitième siècle*, 2 (1914), 232–57, 3 (1915–16), 348–67.

Bouvard, Alexis, 1767–1843
Charles Philippe, 'Notice sur l'astronome Bouvard'. *Revue savoisienne*, 34 (1893), 152–68, 214–25, 285–301. Fragments of autobiography.

Brisson, Mathurin-Jacques, 1723–1806
MS 'Principaux étapes de la vie de M. Brisson, écrits par lui-même'. Library of the Institut de France, Fonds Delambre, 2041/90.

Broussonet, Pierre Marie Auguste, 1761–1807.
MS. written 10 messidor au III/28 June 1795: his life from 1789: Archives Nationales, Paris, F[7] 5142, fol. 45.

Chaptal, Jean Antoine, 1756–1832
Mes souvenirs sur Napoleón (Paris, 1893), 2 vols. Vol. I is Chaptal's record of his own life to 1804; vol. II deals with his recollections of Napoleon.

Cuvier, Georges, 1769–1882
MS. 2598(3) Library of the Institut de France, Fonds Flourens. 'Mémoires pour servir à celui qui fera mon éloge: écrits au crayon dans ma voiture pendant mes courses en 1822 et 1823: cependant les dates sont prises sur des pieces authentiques'. Many deletions by Mme Cuvier.
Printed in Pierre Flourens, *Recueil des éloges historiques* (Paris, 1856), 2 vols. I, 169–93.

Dupont de Nemours, Pierre Samuel, 1739–1817
L'enfance et la jeunesse de Dupont de Nemours raconteés par lui-meme (Paris: Plon-Nourrit, 1906).

Fourcroy, Antoine François, 1755–1809
'Note autobiographique', personal possession of D. Duveen. Quoted in G. Kersaint, 'Antoine-François Fourcroy: 1755–1809: sa vie et son oeuvre'. *Mémoires du Muséum National d'Histoire Naturelle*, série D, 2 (1966), 1–296, passim; p. 242 discusses provenance of the MS.

Hassenfratz, Jean Henri, 1755–1827
G. Laurent, 'Une mémoire historique du chimiste Hassenfratz'. *Annales historiques de la révolution française I* (1924), 163–64. Written 1794.

Haüy, René-Just, 1743–1822
MS. 'Etat de services' (1809). Library of the University of Paris (Sorbonne) MS 1643.
Printed in A. Lacroix, 'La vie et l'oeuvre de l'Abbé René-Just Haüy'. *Bulletin de la Société française de minéralogie*, 67 (1944), 15–16.

Lacépède, Bernard Germain Etienne de la Ville, Comte de, 1756–1825
MS. Fonds Goswin de Stassart, Académie royale des sciences de Belgique. Completed in 1815.
This version printed in R. Hahn, 'L'autobiographie de Lacépède retrouvée'. *Dix-huitième siècle*, 7 (1975), 49–85.
Another version in the Fonds Cuvier, 3209, Library of the Institut de France, utilised for an éloge: Georges Cuvier, *Eloges historiques* (Paris and Strasbourg, 1819–23), 3 vols. III, 285 comments on the text.

Lalande, Joseph-Jérome, 1732–1807
Extracts from MS. autobiography quoted in Comtesse de Salm, 'Eloge historique de M. de Lalande'. *Magazin Encyclopédique*, 2 (1810), 282–325.

Lavoisier, Antoine-Laurent, 1743–1794
MS. 'Notice de ce que Lavoisier, cy-devant commissaire de la Trésorerie nationale, de la ci-devant Académie des sciences, membre du bureau de Consultation des arts et métiers, cultivateur dans le district de Blois . . . a fait pour la révolution'. Printed in E. Grimaux, *Lavoisier* (Paris, 1888), 387–88.

Montucla, Jean-Etienne, 1725–99
Autobiographical letter in G. Sarton, 'Documents nouveaux concernant Lagrange'. *Revue d'histoire des sciences*, 3 (1950–1951), 129–32, dated 21 November 1794, in Archives Nationales, Paris, D. XXXVIII.IV.62.

Moreau de la Sarthe, Louis-Jacques, 1771–1826
Pierre Delaunay, 'Moreau de la Sarthe et ses souvenirs'. *Société française de l'histoire de la médecine*, 30 (1936), 353–62; 31 (1937), 13–42.

162

Périer, Jacques-Constantin, 1742–1818
MS. Autobiography, Bibliothèque historique de la ville de Paris, Nouvelle acquisition 147
 f. 465.

Ramond, Louis François Elizabeth de Carbonnières, 1755–1827
H. Dehérain (ed.), 'Une autobiographie de Baron Ramond'. *Journal des savants*, March
 1905, 121–29. Written in February 1827. The MS. is Fonds Cuvier, 3154 Library of the
 Institut de France.

Sage, Balthasar Georges, 1740–1824
Notice autobiographique, Paris 1818, published by the author.

St-Pierre, Jacques Henri Bernadin de, 1737–1814
Lt-Col. Largemain, 'Bernadin de St-Pierre'. *Revue de l'histoire littéraire de la France* 12
 (1905), 668–92, reprints autobiography.

Tenon, Jacques-René, 1724–1816
MS. Autobiography in Fonds Cuvier 3196, Library of the Institut de France.

Valmont de Bomare, Jacques Christophe, d. 1807.
E. T. Hamy, 'Une autobiographie inédite de Valmont de Bomare'. *Bulletin du Muséum
 d'Histoire Naturelle* 12 (1906), 4–7.

8

NEW SPACES IN
NATURAL HISTORY

From: N. Jardine, J. A. Secord and E. C. Spary (eds.), *Cultures of Natural History* (Cambridge and New York, 1996), 249–265. © Cambridge University Press. Reprinted with permission.

Natural history in transition, 1780–1830

Between the eighteenth and nineteenth centuries natural history underwent profound transformations. An overwhelming interest in evolving classifi-cation systems for specimens of plants and animals was slowly edged out, though never completely replaced, by investigation into aspects of the inward functioning of their physiological systems. 'Natural history' itself slowly sep-arated into separate sub-disciplines such as physiology or palaeontology, each with their own methods, agendas, and subject-matter. At the same time, natu-ral history began to separate from theology, especially in continental Europe, though at a slower pace than in Britain. By the early nineteenth century active men of science began to see natural history as distinct from attempts to argue from the nature of the created world to belief in and knowledge of a benevo-lent deity.

During the same period natural history as a profession also underwent great changes. Much new knowledge about nature continued to be produced by gentlemanly and gentlewomanly amateurs working in largely domestic envi-ronments, supported by an extensive network of clubs and societies. But at the same time, especially in Europe, natural history was increasingly given the backing of major new state-funded and state-controlled institutions staffed by paid full-time expert researchers. Such, for example, were the Natural His-tory Department of the British Museum in London, the Zoological Gardens in London, natural history museums such as that founded by Charles Willson Peale in Philadelphia, or the geological and zoological museums of the new University of Berlin and, above all, the Muséum National d'Histoire Naturelle in Paris, refounded in 1793 on the basis of the former royal botanical gardens or Jardin des Plantes. It is upon this key institution that much of this essay will focus.

DOI: 10.4324/9781003038085-9

The Muséum National d'Histoire Naturelle

The Muséum was undoubtedly the best-known institution in this field in at least the first half of the nineteenth century, attracting many researchers from outside France, and being the home (literally as well as professionally) of many of the leading names in natural history in this period, of men like Georges Cuvier (1769–1832), Jean-Baptiste Lamarck (1744–1829), and Etienne and Isidore Geoffroy St.-Hilaire (1772–1844 and 1805–61 respectively). The Muséum, whose grounds also contained a zoo, a publicly accessible botanical garden, lecture theatres, display galleries, a library, and preparation and dissection rooms, lay on a twenty-acre site on the south bank of the Seine. Throughout this period the site expanded, until it reached almost the same size as the institution occupies today. Unlike today's Muséum, however, the Muséum in the early nineteenth century was not simply a site of scientific labour and of public resort: it was also a domestic space which housed the professors and their families, as well as those of many of the technical assistants and the non-tenured professional staff. Often, their living space was shared between several individuals, and there was little separation between domestic space and working space. Cuvier, for example, reconstructed his own living space so that, by opening a door, he could walk straight from his apartment into the anatomy galleries.

The boundaries of the Muséum enclosed a highly complex space which contained many ambiguous and contested elements. It was a public space which also contained domestic space. It was an open, park-like space in the middle of one of Europe's largest urban centres. It was a space ostensibly dedicated to professional natural history at a time when such paid careers were monopolized by men, and yet was also a space lived in by both genders and all age groups. It was a 'professional space' which was also open to the amateur or even merely slightly interested public. It was a place which focused professional labour on one site, and yet from that site, field naturalists were sent out to wild and inaccessible parts of the world to collect otherwise unknown flora and fauna. The 'open' spaces of the gardens and zoo were flanked by the 'closed' spaces of the galleries and lecture theatres; and each sort of space carried with it a different facet of natural history practice and organization. The Muséum's space was a microcosm which mirrored the debates over the different uses of space which were crucial, in this period of transition, for the future direction of the discipline. Cuvier, for example, expanded the space allotted to his collections in the anatomy galleries, at the expense of other professors, notably Lamarck, with whose classification systems he disagreed. He and his brother Frédéric, who directed the zoo in the Muséum, clashed over allocation of extra space between the living animals and the anatomy galleries, a clash which mirrored their debate over the importance of work on the living animal, or the dead specimen divorced from habitat.

The Muséum was a space where not only contestations about classification could take place, but also one where entirely new approaches to nature could be worked out. Naturalists like Cuvier were creating the new science of comparative anatomy, by which it was hoped to generate a new way of looking at the relationships between living beings. Cuvier tried to relate living beings to each other by comparing their internal structures rather than their external characteristics. For museum naturalists often working from imperfectly preserved specimens, this could have two consequences. Firstly, an emphasis on detailed anatomical exploration of internal physiological structures, to discover the particular ways in which systems related, and to use those relationships as the basis for grouping living beings together. This way of classifying tended to privilege function over form. Secondly, the expansion of comparative anatomy also enabled much exploration in palaeontology to take place. Fossil remains were being discovered in Paris in great numbers from the 1790s, due to excavations for the expanding city planned by Napoleon. Naturalists like Cuvier also devoted much time and energy to obtaining careful copies or drawings of specimens of fossils known elsewhere. As French conquests in Europe proceeded, newly conquered territories contributed to the collections of the Muséum. It was in this way that complete skeletons, such as the Madrid *Megatherium*, became known in Paris. All this meant a change in the nature of the Muséum space. In the later eighteenth century the Muséum had been largely a botanical *garden*; after 1793 it began to be an institution which remained an outdoor setting for the display of nature but also began to house increasing numbers of built spaces like galleries and dissection rooms to house an increasingly indoor science. This also put great pressure on the vocational ideals of naturalists. Where was their science located? Indoors or out? Were the systems of explanation created by the work of indoor anatomists superior to the intimate knowledge of living creatures in their habitats which was traditional field natural history?

The meaning of space

Recently, the relation between spatial disposition and intellectual authority has become a new focus of study in the history of science and beyond. Spatial metaphors abound in recent philosophical works. Charles Taylor, for example, in an important recent work, *Sources of the Self* (1989) posits that the basis of the self is a 'moral space', from which the self speaks. Alasdair MacIntyre in *After Virtue* (1985) argues that rational enquiry has intentions that are only intelligible in virtue of their settings. Both would agree that what passes as rational belief has not merely a history, but a geography as well. Spatial metaphors in studies of the distribution of scientific influence, in the transmission of scientific instrumentation, and the disposition of space in 'laboratory life' have been investigated in well-known works.[1] The 'built spaces' of science, especially the anatomy theatres and public lecture rooms, as well as institutions of popular science such as mechanics' institutes, have been investigated extensively by architectural historians. Such buildings, as has been shown, both manifest and impose structures of authority between teachers and taught, expert and public. The design of buildings can also both pressure and structure the access to knowledge of different social classes. Historians have long known that, in a broader sense, science varied with location. The styles of science are regionally differentiated, as is the reception of scientific themes. Sociologists like Erving Goffman have pointed out that the spaces of scientific knowledge have often determined the degree of credence given to claims to expert knowledge. This is a point also made in Steven Shapin and Simon Schaffer, *Leviathan and the Air Pump* (1985).

On the other hand, little attention has been paid to the 'spaces' of science which lie outside the built environment, spaces like the public botanical and zoological gardens of the Muséum, or indeed the whole area of 'wild nature' into which its 'field' naturalists ventured to find specimens for examination by the experts of Paris. How did they see 'the wild' or 'the field'? Also, little attention has been paid to the use of 'domestic' space in science. What were the interactions *between* these very different sorts of spaces? In the confines of this chapter we cannot hope to go more than a little way towards resolving these questions, but we can at least open them up. We can begin to ask more fundamental questions about how contemporaries categorized and experienced different forms of space: we can also explore our own presuppositions about what space is.

'Space', after all, is a profoundly ambiguous concept. The word itself is unclear. It can mean either 'the area contained within a given boundary', or it can mean the opposite, unbounded area, as in the expression 'outer space'. Phrases like 'space, the final frontier', resonate as they do precisely because they encapsulate the contradiction inherent in our idea of space. Secondly, it has also been the case that spatial perception has been treated very largely as equivalent to *visual* experience alone. However, much recent discussion has been redefining spatial perception as an experience in which the *whole* body is implicated.[2] In other words, to reconstruct the spatial experience of the natural historians of this period we have to think about not just what they might have seen, but also about what sort of psychological structures mediated their response to space. It is not possible to explore space as if it were a simple or autonomous entity, without questioning the inner world of its human perceivers. To find out what space perception was, we have to come to grips with the whole-body experience of the men of the past. Their 'inner-space' also structured what spaces were seen, and how.

'Space', therefore, can mean many different things. It can mean the bounded, built spaces in which natural history located itself. It could also, especially in natural history 'fieldwork', be experienced as 'movement over' terrain, which allowed the naturalist to observe a rapidly changing population of living beings in their habitat. In this kind of natural history, knowledge-gathering was inseparable from movement *through* space, inseparable therefore from bodily involvement. In this way, natural history fieldwork was squarely associated with a tradition which linked curiosity to movement.[3] Conversely, other kinds of natural history which concentrated on the dissection of specimens collected by field natural historians, carried out in bounded, built environments such as anatomy rooms, carried quite a different epistemological tradition.

Different spaces too had different social and political connotations. The Muséum was also, as we have already noted, a public space, a place to which the ordinary population of Paris was able to resort freely, where it could wander among the gardens and admire the live animals in the zoological gardens, or, indoors, take in the wonders displayed in the galleries devoted to displays of the dead specimen. It was a place for gentle exercise and refined entertainment in the heart of the metropolis. In 1800, the Muséum's professor of chemistry, Antoine Fourcroy, described the gardens of the Muséum as an 'Elysium', or abode of the gods.[4] Fourcroy's description should alert us to the extent to which the Muséum's character as a 'public' space was also dependent on its definition as a space which could be assimilated to the mythological structures of classical antiquity, and could thereby be conceptualized as a space outside space: a heavenly place off the grid of real-world maps, and hence outside the focus of social and political production. This way of conceptualizing the Muséum was reinforced by the way in which its publicly accessible collections brought together in one place the whole range of the natural order, which, in the 'real world', would never be found together in one space. The rapidly growing palaeontological galleries also meant that the visitor to the Muséum could see not only the denizens of many different parts of the earth's surface together in one spot, but also the products of many different *eras* of the earth's history. In thus making visible nature from all times and from all spaces, the Muséum intensified its character as an ideal place, a place where all other times and places were gathered, a claim which remained extremely powerful because it was supported by the spatial layout of the Muséum itself. Thus the Muséum could remain unaffected, for the ordinary public, by the quarrels about the order of nature which caused such conflict amongst its professional staff.[5]

This 'ideal' character of Muséum space was heightened by its location in the heart of a great capital city. The Muséum was resolutely divorced from its congested, built-up setting, in a part of Paris which was undergoing considerable 'development' in the early nineteenth century. Yet paradoxically this very setting acted to emphasize, by its contrast, the Muséum's character as an ideal space, not generated by other surrounding spaces. This ideal character at once obscured, and at a profound level mirrored the political conflict of the French Revolution in which the Muséum itself lay embedded. The idea of a botanic garden as an Eden or an earthly Utopia long antedated the foundation of the Muséum in its modern form in 1793. But the cultural impact of the idea was to become far greater, when the Muséum became accessible to the general public, and was notionally owned by the nation which claimed to represent that general public. In the case of the Muséum, this change in relation to the larger public was particularly dramatic, and shows the interaction between the growth of representative politics, and the increase in public consumption of natural history.

At the height of the French Revolution, in 1793, the Muséum was given its own constitution, which entailed the transformation of the former royal Jardin des Plantes into a national possession, designated as a place of public resort and instruction in natural history. At a profound level, its transformation was in line with the violent transformation of France from monarchy to republic, from dynastic state to representative republican government. Its transformation was also congruent with the Jacobin rhetoric which erected 'nature' and the 'natural' almost into ethical norms. The Muséum, in this revolutionary mode, was supposed to contribute to the moral, and hence political, transformation of Frenchmen into citizens of the Republic of virtue, by rendering the order of nature visible, accessible, and transparent to the sovereign people, who had replaced the sovereign monarch.

In reality, the Muséum owed much to violence, and was characterized by internal strife. Its collections were hugely amplified by the assault of Revolutionary terror on the aristocracy, and later by the victories of Revolutionary armies over the monarchs of Europe, which allowed their 'liberated' collections to be transported to the Paris Elysium. Muséum staff frequently quarrelled over what the order of nature actually was, resulting in the display of mutually incompatible classification systems in the public anatomy galleries. The 'space out of space' of the Muséum also often had to be brutally constructed, put together with a sledge-hammer *bricolage*. Cuvier, for example, described in his autobiography the typically direct way in which he extended his own display space, by breaking through from the Muséum buildings into another adjacent building left vacant by the demise of an old-regime administrative body.

But it was the same Cuvier who in 1819 was to describe the Muséum public gardens as 'a temple', as 'the largest and most beautiful ever consecrated'.[6] Why was there such investment in the 'ideal' of the gardens, in their image as an earthly paradise, where the public could find moral uplift and civic regeneration by contact with the ordered display of nature, even by men who well knew the conflict and chaos which had gone into the making of those particular Elysian fields?

Partly this was because an image of a public space outside social production, fitted in so well with the new ideas of the collectivity, the 'nation', which was the political innovation of the Revolution itself. In 1793 military crisis on the frontier, and civil war within France led the Jacobin politicians in power to put even greater stress on the collectivity which they called the 'nation'. The 'nation' was supposed to replace the old-regime social order, which divided groups and regions from each other by privilege. The 'nation' on the other hand, was supposed in theory to be a collectivity of individuals equal before the law, without class, gender, or political conflict. The political realities of the Revolution and succeeding periods were far from reflecting this ideal. Just as the social fabric was now to be conceived as seamless, without social forces or conflicts, so too the Muséum was to be conceived as a 'Utopian' space equally removed from conflict, whose visitors would not see reflected in front of them the fragmented world of actuality, but a seamless order of nature.

This idea of the gardens of the Muséum as an Elysium also fits in with the famous analysis by the social theorist Walter Benjamin, in which he likened such Utopian spaces as parks to a collective dream:

In the dream in which every epoch sees in images the epoch which is to succeed it, the latter appears coupled with elements of pre-history, that

is to say, a classless society . . . to give birth to the utopias which leave
their traces in a thousand configurations of life from permanent buildings
to ephemeral fashions.[7]

In exactly the same way, the Muséum offered the Paris public access to a space
outside space, a space which refused to mirror the social and political conflicts
so harshly mapped on to the surrounding social world. The Utopian space of the
Muséum also corresponded to the essential characteristics of the political categories such as 'the nation' or 'the general will' or 'the sovereign people' which
had arisen from the Revolution. This is an important point to make, because
Benjamin associates the rise of Utopian public space with the onset of industrialization. The example of the Muséum seems to show, rather, that Utopian
public space is generated by the enormous shift in political categories which
accompanied the French Revolution, and thus long preceded large-scale industrialization in France.

Field natural history and wide open spaces

The public spaces inside the Muséum site thus associated it, and thereby natural history, with a Utopian and panoramic view of nature. The Muséum's connection with 'field' natural history was also associated with the production of
images of the practice of natural history which were heroic and exemplary.
Field naturalists, whether dominant and heroic figures like Alexander von
Humboldt, or the less well-known figures sent out by the Muséum as '*natu-
ralistes-voyageurs*', were imagined by a growing public for popular science
as struggling over remote and dangerous terrains in dedicated pursuit of new
and strange plants and animals. Public veneration of such figures, who came
closer than any other men of science to emulating the heroic men of action
so central to nineteenth-century European imperialist mythology, however,
masked an increasing struggle within natural history over the value of field
natural history.

The concept of 'the field' itself is a complex one, and we must await its definitive exploration by historians. But it is nonetheless important for our purposes,
because the idea of 'the field' is pivotal in its union of spatial metaphor and epistemological assumptions.

Field natural history was closely associated with a particular approach to
nature, as well as with ideals of heroic, manly endeavour. Increasingly, however, the field naturalists, and the naturalists who theorized about the order
of nature, or who conducted wide-ranging comparative research in taxonomy,
physiology, or anatomy, usually within built environments, ceased to be the
same people. Charles Darwin, who published his *Origin of Species* in 1859,
the year of Alexander von Humboldt's death, was probably the last man of
science who combined all these different aspects of natural history in his
own life.

Earlier in the century the battlelines had already been drawn. In 1807 the Muséum professor Georges Cuvier reviewed a report of field research by the scientific hero Humboldt. He commented:

> Usually, there is as much difference between the style and ideas of the field naturalist ('*naturaliste-voyageur*'), and those of the sedentary naturalist, as there is between their talents and qualities. The field naturalist passes through, at greater or lesser speed, a great number of different areas, and is struck, one after the other, by a great number of interesting objects and living things. He observes them in their natural surroundings, in relationship to their environment, and in the full vigour of life and activity. But he can only give a few instants of time to each of them, time which he often cannot prolong as long as he would like. He is thus deprived of the possibility of comparing each being with those like it, of rigorously describing its characteristics, and is often deprived even of books which would tell him who had seen the same thing before him. Thus his observations are broken and fleeting, even if he possesses not only the courage and energy which are necessary for this kind of life, but also the most reliable memory, as well as the high intelligence necessary rapidly to grasp the relationships between apparently distant things. The sedentary naturalist, it is true, only knows living beings from distant countries through reported information subject to greater or lesser degrees of error, and through samples which have suffered greater or lesser degrees of damage. The great scenery of nature cannot be experienced by him with the same vivid intensity as it can by those who witness it at first hand. A thousand little things escape him about the habits and customs of living things which would have struck him if he had been on the spot. Yet these drawbacks have also their corresponding compensations. If the sedentary naturalist does not see nature in action, he can yet survey all her products spread before him. He can compare them with each other as often as is necessary to reach reliable conclusions. He chooses and defines his own problems; he can examine them at his leisure. He can bring together the relevant facts from anywhere he needs to. The traveller can only travel one road; it is only really in one's study (*cabinet*) that one can roam freely throughout the universe, and for that, a different sort of courage is needed, courage which comes from unlimited devotion to the truth, courage which does not allow its possessor to leave a subject until, by observation, by a wide range of knowledge, and connected thought, he has illuminated it with every ray of light possible in a given state of knowledge.[8]

This passage is worth quoting at some length because it resumés so many connected issues. The sharp division it draws between two ways of doing natural history is organized by the contrast between two different experiences of space, which entail two contradictory ways of approaching nature. The field naturalist has unbounded spaces at his disposal. His observations are vivid, instantaneous, active, and dramatic, and allow him to display all the manly qualities of courage, energy, and mastery of the earth's surface. But he has little *over*view of the natural order as a whole; his view of individual beings is fragmented and insecure, in spite of their momentary precision and vividness.

In asserting that the only real voyages of exploration into nature take place in the sedentary naturalist's study, Cuvier is thus making a claim that mastery over and comprehension of nature come not from *passage over* terrain, but from the steady and immobile *gaze* of the sedentary naturalist. Courage means the courage to look steadily: look in the face of Medusa, he seems to be saying. He is thus also saying that true knowledge of the order of nature comes not from the whole-body experience of crossing the terrain, but from the very fact of the observer's *distance* from the actuality of nature. True observation of nature depends on *not* being there, on being anywhere which is an elsewhere. At bottom, Cuvier is fighting an epistemological battle. If truth is space-dependent, then the way is left open for unlimited relativism. The claims of natural history to be 'true' may thus, paradoxically, be damaged by the very fieldwork seen as central to it by the public perception of science. Such an argument is entirely consistent with Cuvier's lifelong struggle to raise the true status of natural history. It is also noticeable that Cuvier is having to make this argument with considerable force, to an audience he supposes will be hostile to, or at least unfamiliar with, these ideas. He is still having to argue against the 'culture hero' of the field naturalist, one who was prepared to pay with his own person, with rough travel, solitude, isolation and danger, for his insights, that the sedentary study of nature was not unmanly.[9] It was a conflict which Cuvier's interest was not to settle, but rather to sharpen. For the rest of the century, field naturalists were often to resist theory and systems on the grounds of avoiding over-generalization.

Cuvier also has to travel descriptively between the inner and outer worlds of the two types of naturalist he is describing. He is well aware that the different relationship to external space manifested by each sort of naturalist is inseparable from a different structure of 'inner space'. The field naturalist is seen as highly responsive, engaging with each passing incident in the natural world around him, erecting few or no defences against the passage of rapid, immediate impressions into his inner world. The sedentary naturalist, on the other hand, is seen as preoccupied with both physical and psychic distance, and with the belief that out of distance comes truth. In doing this, Cuvier is engaging natural history with a new cultural value.

Intellectual historians have recently pointed out how the very concept of psychical 'distance' – as in phrases such as 'he gave a distant stare' or 'a distant manner' – starts to be elaborated in the eighteenth century, at exactly the same time as aesthetics begins to praise landscape painting based on a perspective projection into far distances. At the same time, new techniques of instrumentation were allowing distances, especially in the heavens and at sea, to be measured with new degrees of accuracy.[10]

'Distance' was a new cultural value which affected, and brought together under a single lens, both information and perception. Field naturalists were validated by their heroism in physically encountering and overcoming distance from metropolitan centres. Conversely, sedentary naturalists were forced to argue that their *psychic* distance from the object of their study guaranteed the superior truth-value of their brand of natural history. It was not a big step from the establishment of distance as a cultural value over which it made sense to struggle, to the production of the idea of objectivity, meaning precisely the placing of 'distance' between the observer and the observed, between the knower and his own responses. For the eye of the field naturalist, seduced by the dazzle of passing events, Cuvier's sedentary naturalist substitutes an observation which is distanced, and thereby dominating in its control over the whole range of the natural order. It was in this way that a previous era had often conceived of the eyes of princes moving over the conspectus of their own territories, just as the sedentary naturalist passed all nature 'in review' before him, and through this gaze controlling it, rather than running the risk of being overwhelmed by it.[11]

The naturalist and inner spaces

The 'inner space' of the sedentary naturalist, with its high degree of compartmentalization between observation and response, inner and outer worlds, comes very close to exemplifying what Norbert Elias, in his groundbreaking study of the history of emotional structure, has called '*homo clausus*' (enclosed man). By this, he refers to a human type which he sees as emerging from courtly society between the Renaissance and the end of the eighteenth century, one whose identity was based precisely on the maintenance of his own 'distance', emotional and physical, from other human beings, with a strong consciousness of himself as separate, walled off from others by invisible barriers. With the emergence of this human model came restrictions on the exploration of the world through the whole body. Conduct books, for example, recommended children to understand new things by looking but not by touching. *Homo clausus*, just like Cuvier's sedentary naturalist, was supposed to refuse access to his closely guarded inner space to the disorderly rush of sensation, the bodily effort and response, which accompanied field natural history. In contrast, Muséum work in natural history needed close physical control involved in the preparation of delicate specimens. Correct assessment of specimens depended on a highly controlled orderly environment which screened out the inessential or the background of detail.

I have argued elsewhere that *homo clausus* is also the human type on which the new political world of the French Revolution was to be based, the model for the sort of human being who was an autonomous, and thereby legitimate, political actor, and hence was also the privileged location of 'rights'.[12] It is no accident, therefore, that natural history should also begin to construct its own 'ideal person' in a different way from the past, and that that construction should involve the same conflict between inner and outer spaces and be as deeply influenced by the impact of the new political values. Just as the Utopian 'public spaces' of the Muséum were constructed at the height of Revolutionary terror, and were congruent with the Revolution's Utopian, conflict-free idea of 'the nation' so too was the replacement of the responsive inner space of the roving field naturalist by the controlling eye of the confined sedentary enquirer.

'Curiosity' had indeed been replaced by a crisis and conflict over definitions of both inner and outer spaces in natural history, a crisis that was to provide obstacles to the maintenance of natural history as a unified discipline.

Notes

1 Sophie Forgan, 'Context, image and function: a preliminary enquiry into the architecture of scientific societies', *British Journal for the History of Science,* 19 (1986), pp. 89–113; 'The architecture of science and the idea of a university', *Studies in History and Philosophy of Science,* 20 (1989), pp. 405–34; Thomas A. Markus, *Buildings and Power: Freedom and Control in the Origins of Modern Building Types* (London and New York, 1993), pt. 3, 'Buildings and knowledge'. For some exploration of domestic space in science see P. Abir-Am and D. Outram, *Uneasy Careers and Intimate Lives* (New Brunswick, 1987).

2 M. Milner, *Eternity's Sunrise* (London, 1987); see also the psychoanalytical theory discussed in Klaus Theleweit, *Male Fantasies,* 2 vols. (Oxford, 1987); Henri Lefebvre, *The Production of Space* (Oxford, 1991).

3 K. Zacher, *Curiosity and Pilgrimage* (Baltimore, 1976).

4 'These gardens are an Elysium, which the friend of nature cannot approach without awe', quoted in D. Outram, *Georges Cuvier,* p. 161.

5 See Outram, *Georges Cuvier,* pp. 161ff., on the way the arrangement of the galleries represented conflicts over classification systems, and themselves represented an effort to *display* spatially the structure of nature which rendered them very different from early modern 'cabinets'.

6 Outram, *Georges Cuvier,* pp. 176, 251; Georges Cuvier, *Recueil des Éloges Historiques,* 3 vols. (Paris and Strasbourg, 1819–27), vol. III, p. 436.

7 Walter Benjamin, 'Paris, capital of the nineteenth century', in C. Hoare (ed.), *Charles Baudelaire: A Lyric Poet in the Era of High Capitalism* (London, 1973), pp. 155–76, p. 159.

8 Quoted in Outram, *Georges Cuvier,* pp. 62–3.

9 For a forceful description of field natural history writing, see Larzar Ziff, *Writing in the New Nation: Prose, Print and Politics in the Early United States* (New Haven, 1991), pp. 34–54, which examines field natural history writing as a way of controlling and defining terrain. The connection between manliness and field science is stressed in J. Secord, 'The Geological Survey of Great Britain as a research school, 1839–1855', *History of Science,* 24 (1988), pp. 223–75; Michael Shortland, 'Darkness visible: underground culture in the Golden Age of geology', *History of Science,* 32 (1994), pp. 1–61.

10 J. L. Jarret, 'On psychical distance', *Person,* 42 (1971), pp. 61–9; M. H. Nicolson, *Mountain Gloom and Mountain Glory: The Development of the Aesthetics of the Infinite* (New York, 1963).

11 L. Marin, *Portrait of the King* (London, 1988); Jacques Revel, 'Knowledge of the territory', *Science in Context,* 4 (1991), pp. 133–61. The literature on objectivity is vast; see Zeno G. Swijtink, 'The objectification of observation', in L. Kruger, L. Daston, and M. Heidelberger (eds.), *The Probabilistic Revolution* (Cambridge, MA, 1987), vol. I, pp. 261–86.

12 D. Outram, *The Body and the French Revolution* (New Haven and London, 1989); Norbert Elias, *The Civilising Process,* 2 vols. (New York, 1978); Jean de la Salle, *Traité de la civilité* (Paris, 1774), p. 23: 'Children like to touch clothes and other things that please them with their hands. This urge must be corrected, and they must be taught to touch all that they see only with their eyes.'

Further reading

Appel, T., *The Cuvier–Geoffroy Debate* (Oxford, 1987).

Atran, S., *Cognitive Foundations of Natural History: Towards an Anthropology of Science* (Cambridge and Paris, 1990).

Corsi, P., *The Age of Lamarck* (Berkeley and Los Angeles, 1988).

Foucault, M., 'Knowledge, space, power', in P. Rabinow (ed.), *The Foucault Reader* (London, 1986), pp. 239–56.

Lefebvre, H., *The Production of Space* (Oxford, 1981).

Limoges, C., 'The development of the Muséum d'Histoire Naturelle of Paris, 1800–1914', in R. Fox and G. Weisz (eds.), *The Organisation of Science and Technology in France, 1808–1914* (Cambridge and Paris, 1980).

Ophir, A. and Shapin, S., 'The place of knowledge: a methodological survey', *Science in Context,* 4 (1991), pp. 3–21.

Outram, D., *Georges Cuvier: Vocation, Science and Authority in Post-Revolutionary France* (Manchester, 1984).

———, 'Uncertain legislator: Georges Cuvier in his intellectual context', *Journal of the History of Biology,* 19 (1986), pp. 328–68.

———, *The Body and the French Revolution* (New Haven, 1989).

9

ON BEING PERSEUS

New knowledge, dislocation,
and enlightenment exploration

From: David N. Livingstone and Charles W. J. Withers (eds.), *Geography and Enlightenment* (Chicago and London: Chicago University Press, 1999), 281–294. © 1999 by The University of Chicago.

The eighteenth century believed perhaps more strongly than any other that travel makes truth. The problems raised by such a belief in an era which was both that of the Enlightenment and that of the exploration and mapping of that one-third of the earth's surface which is the Pacific ocean are the subject of this paper. The link between traveling and truth finding sits deep in the organizing mythologies of the West, whose origins in Greece and Rome were fully familiar to the Enlightenment and provided much of the basis for its self-fashioning. One of the most powerful of such myths is that of the Greek hero and wanderer, Perseus.

In Greek mythology, Perseus confronted the three-headed monster called the Gorgon. He cut off the monster's only mortal head, Medusa, whose head was covered with snakes and whose gaze turned to stone all who encountered it. In order to do this, Perseus made the monster confront her own reflection in his burnished shield. Once she was petrified by her own gaze, Perseus was able to lop off her head. We are told too that after this, Perseus returned to his own kingdom and found there his hall full of rival claimants to his throne. When he held up his terrible trophy on the threshold, the head of Medusa stared down the assembled throng and turned them to stone, frozen in mid-breath, in the very act of perceiving their danger. Perseus then presented Athena, goddess of clear sight, of rationality, with Medusa's head, which trophy this armed goddess hung henceforward on her own shield.

Even in this bare and incomplete outline, this is a myth that tells us much. It tells us about the power and the terror of sight. It tells us about the transformation of life into the frozen forms of death. It also tells us much about great journeys and their ending, for Perseus travels to the westernmost extremities of the earth to reach the Gorgon. And it tells us much about the costs of such travel. For Perseus, in many versions of the tale, reaches the Gorgon's lair after many prior ordeals, in the depths of fatigue – reaches it not in a hero's poised certainty of victory, but with his courage already failing, already about to freeze in the held breath of his own risk and terror. This is why, at the last moment, one version of the story runs, his courage failed him. Turning his head away, he averted his eye from Medusa and relied on the goddess Athena

DOI: 10.4324/9781003038085-10

to guide his arm in the fatal blow. Thus the myth of Perseus is also a myth about the relationship between heroism and vulnerability. It is also a myth about the interpolation of action, terror, and, that supremely Enlightenment category, rationality.[1]

This is a disturbing myth. There is so much of the terror, so much of the devastation of seeing, in such close companionship with such rationality and such power. By thinking about the myth of Perseus in the shape of this unease, we begin also to enter some of the unease with which explorers' actions and the specific forms of rationality associated with the knowledge they produced also affected the Enlightenment. But to talk about unease in this context may seem strange. The Enlightenment has often been presented precisely as an heroic age, one which saw the emergence, and triumphant success, of exploration conducted primarily for scientific ends. We can, after all, scarcely imagine the map of the world without the knowledge produced by the Pacific and Antarctic explorations of this period, or the Enlightenment itself without the knowledge produced through the journeys of such men as Cook, Pallas, Bougainville, La Condamine, or Alexander von Humboldt. The imagination of the Enlightenment is incomplete without the new plants and animals, minerals, fossils, and geographies recorded and brought back by explorers. At an even deeper level, what explorers reported about the new human societies they encountered made the whole issue of human difference itself a central problem in the Enlightenment. For if there was one human nature, why then did human societies differ so profoundly? If there was one human history, and one human rationality, then how could the human social world be ordered with such absolute difference in its particular manifestations?

And yet, explorers' knowledge, an interpolation of rationality and action, was also a serious and disturbing problem, in ways specific to the Enlightenment. What sort of knowledge, after all, was explorers' knowledge? What secured its status as truth? The testimony of the explorer himself was, by itself, no guarantee of anything, when such testimony precisely concerned the distant, the new and the previously unseen. Such knowledge could not be experimentally replicated and was not produced in front of an impartial and freely assembled audience, or in the context of any of the adjudicating institutions of European science. Explorers, it is true, often returned with artifacts and natural specimens from the distant lands they visited. But artifacts could bear witness only to themselves. They could by themselves prove nothing about the truth status of the narrative of discovery and exploration in which they were alleged to lie embedded.[2] Explorers' knowledge thus posed in stark form the problems of belief, trust, and facticity.

Trust has become a fashionable theme in recent writing in the history of science. The essence of the explorer's claim was to be trusted as an eyewitness to a world that few or no others had seen. Yet it was precisely here that the explorer ran straight into all the Enlightenment's problems with accepting such testimony as compelling belief. These were the problems which were discussed in David Hume's 1748 *Essay on Miracles*, or Voltaire's 1765 *Questions sur les miracles* (see also Shapin 1994). Eyewitness testimony, as Voltaire and Hume demanded, became increasingly measured against the grid of a newly elaborated and powerful idea of probability. The linkage of the exotic, the hitherto unknown, and the marvelous, began to decline (on this point, see Daston 1988; Daston and Park 1998). But even here, explorers' knowledge, precisely insofar as it was knowledge of the new, resisted the grip even of an increasingly powerful probabilism. In the end, it could be accepted only on the basis of trust. Such trust could be built up by means of authorship, and it is not surprising that many explorers, most notoriously Alexander von Humboldt, invested perhaps as much in writing the narrative of travel as they did in that traveling itself. But the trust crucial to turn a traveler's tale into an exploration account, which makes much higher truth claims, could not be produced by authorship alone. The essential and prior step was to place trust in the explorer himself: trust in his moral integrity and trust in his perceptual accuracy. Both were important. The explorer had to be trusted both to see and to tell, had to be trusted to interpolate action with rationality. And such trust was very difficult to command precisely because the philosophy of the Enlightenment called into question this very link between the moral self and the perceiving self of the explorer. Exploration knowledge was a particularly acute form of this problem because so much knowledge produced in exploration depended on the sense impressions of the explorer. Even if he used instruments to extend and calibrate sense impressions, what the explorer himself saw was crucial to establishing the truth-status of his observations. This meant that exploration knowledge was profoundly at odds with attempts since the scientific revolution of the seventeenth century to find epistemological legitimacy for experiment-based science in a denigration of knowledge based on the senses (Dear 1987; Shapin 1988). Just as it was at odds with the Enlightenment's questioning of the necessary veracity of the eyewitness observer.

Enlightenment philosophy also went to some lengths to question the very possibility of the integrity of the knowing subject. A strong sense of the self that knows itself to be a self, and is thus capable of being a locus of moral identity, seemed increasingly for the Enlightenment to be placed in question by the very operations of the senses on which explorers relied to make contributions to knowledge. Hume and Condillac, for example, questioned whether there could be an epistemological unity of consciousness. Hume famously remarked in his 1739 *Treatise on Human Nature*, "I may venture to affirm of mankind, that they are nothing but a bundle or collection of different perceptions, which succeed each other with an inconceivable rapidity, in a perpetual flux and movement." (I, iv, 6) This was a theme pursued by the French philosopher Condillac in his *Traité des Sensations* of 1754. In his work he uses the heuristic device of a living statue to ground his debate on human sense impressions. It is not long before the statue utters the painful words:

> . . . I know that the parts of my body belong to me, but I am unable to understand how. I see myself, I touch myself, in one word I experience myself, but I do not know what I am. And if I believed that I was sound, taste, colour, smell, now I know no longer what I should believe that I am.
>
> (IV, 8, 6)

In other words, personal identity is constitutionally discontinuous. This means that the self which processes sense impressions may have no clear relationship with the moral self. The discontinuity of the self was not a new idea in the Enlightenment. But it was given a new twist by the increasing reluctance of the Enlightenment to find a saving unity in the idea of the soul. Instead, increasing attempts were made to ground identity in sense impressions at exactly the same time as the discontinuity of those impressions was admitted. This has the consequence that sense impressions in fact pull away from the possibility of moral authentication.

Explorers' knowledge might thus seem to be the exemplification of Enlightenment concerns about the difficulty of knowing anything securely about the external world, of making a secure relationship between knowledge and the knower, or of persuading others of the validity of knowledge gathered through ineluctably individual sense impressions. Could the explorer, like Perseus, ever find a way of doing the work of Medusa, of freezing the world into the forms of communicable truth? If one reads what explorers themselves sometimes say about the actuality of their cognition, then it might appear that the philosophers were right, that to perceive the dazzle and the glitter of the world really did pose a threat to a unitary personality capable of moral discipline, capable of being trusted. Let us to this point read a famous letter by Alexander von Humboldt, written in July 1799, a few days after his first landfall in Venezuela with his traveling companion Aimé Bonpland. In a state close to ecstasy, he wrote home to his brother Wilhelm in Berlin:

> What a fantastic and extravagant country we're in! Fantastic plants, electric eels, armadillos, monkeys, parrots, and many real, half-savage Indians . . . and what trees! Cocoa-nut trees fifty or sixty feet high, *poinciana pulcherrima* with a foot-high bouquet of magnificent, bright red flowers . . . a host of trees with enormous leaves and scented flowers, as big as your hand, completely unknown to us. . . . And what colours in birds, fish, even cray-fish, sky-blue and yellow. We rush around like the demented. In the first three days we were quite unable to classify anything, we'd pick up one object only to throw it away for the next. Bonpland keeps telling me that he will go mad if the wonders do not cease soon.[3]

This is a description of the boundary between ecstasy and sensory disorganization. It is a description of falling suddenly into a world where everything was brightly colored and strangely shaped, and where nothing bore any relation to previous experience. Perception was thus totally unshaped by memory. The brilliance of the world, its hallucinatory dazzle, threatens to dissolve those two alleged entities called Humboldt and Bonpland into a cacophony of sense impressions bordering on insanity. And the power of the experience of the exotic is also powerful enough to shatter the aesthetic communities on which the possibility of its description rests. Not only might the explorer himself be dissolved by the intensity of what he saw, he might also be unable to describe what he saw to those who did not. This is why Alexander's letter

from Venezuela to Wilhelm in Berlin walks a fine line between the possi-
bilities and the limits of the description of the new. And this is exactly what
exploration is about: establishing the new as a category of experience.

Everything I have said up to this point has emphasized the problems raised
by exploration knowledge in the Enlightenment. It is not surprising that it
aroused sustained attacks by contemporaries. To a savant of the stature of
the anatomist and taxonomist Georges Cuvier, for example, exploration, like
all field science, was highly suspect. In an 1807 review of Humboldt's *Tableaux de la Nature*, Cuvier forcefully argued against the claims of fieldwork
to produce adequate knowledge of nature through direct observation. He did
so through a direct attack on the cognitive and moral make-up of the field
investigator. For Cuvier, the procedures of what he calls the '*naturaliste-voyageur*'
were irreconcilable with those of the sedentary naturalist. The traveling
observer was doomed to remain precisely that. Cuvier remarks that the field
naturalist passes through ". . . at greater or lesser speed, a great number of different places, and is struck, one after the other, by a great number of interesting objects and living things," but ". . . he can only give to them a few instants
of time and thus his observations are broken and fleeting." This sounds almost
as though Cuvier had read Humboldt's letter of 1799. Cuvier then summons
up the figure of the sedentary naturalist, the man who works in the collections
and museums. He admits that such a man ". . . only knows living beings from
distant countries through reported information subject to greater or lesser
degrees of error . . . the great scenery of nature cannot be experienced by him
with the same vivid intensity as it can by those who witness it at first hand."
And yet, Cuvier argues, directly attacking Humboldt, ". . . if the sedentary
naturalist does not see nature in action, he can yet survey all her products
spread before him. He can compare them with each other as often as is necessary to reach reliable conclusions. He defines his own problems." Cuvier
concludes with the remarks that ". . . the travelling observer can only travel
one road. One can only roam freely through the universe, by staying in one's
study. For that, a different sort of courage is needed . . . courage which does
not allow its possessor to leave a subject, until by observation and . . . connected thought, he has illuminated it with every ray of light possible in a given
epoch of knowledge."[4]

For Cuvier, the formation of new knowledge in continuous passage through
space, the very essence of exploration, was deeply suspect. For him, travel does
not make truth. On the contrary. For Cuvier, mastery over, and real comprehension of the order of nature, comes not from passage but from immobility. The
apparently heroic field observer lies, in fact, under the tyranny of the immediate. Real courage, Cuvier argues, means to create an ordered structure out of the
immediacy of experience, not to be tempted by its momentary vividness. Such an
argument shows one way to discuss location and dislocation in relation to experience and epistemology. Cuvier argues that true knowledge of the order of nature
comes from the very fact of the observer's distance. It depends exactly on not

being there, on being anywhere which is an elsewhere. This was a way of avoiding all the problems which the Enlightenment had debated in relation to sense impressions, to eye-witnessing, and to the assessment of the moral integrity of the distant field observer in continuous transit. It is important that Cuvier does not refer to observation by means of scientific instruments to answer or avoid these questions. Just as much as Humboldt, he assumes that the central question is the assessment of the quality and location of the naturalist himself, not on his use of instruments. His version of 'objectivity' turns rather on the experiential categories of location and dislocation.

Was this an error? Surely instruments offered a golden opportunity to avoid all the problems raised by direct sense impressions of the new and the transitory? It is highly tempting to see instrumental readings as a way of freezing and transporting the alien reality with which the explorer was surrounded. James Cook took a battery of instruments with him to the observation station at Point Venus, on Tahiti, in 1769; Humboldt took a huge collection of the latest devices to Venezuela. The British Arctic expeditions of the 1820s were given specific mandates to test new forms of compasses and atmospheric gauges (Bravo 1992; Licoppe 1996; Schaffer 1988; Sorrenson 1996). We also know that the disciplines of creating a regular series of instrumental observations under difficult conditions were one of the guarantees of an explorer's probity. And from such series previously unsuspected global regularities were detected, such as Humboldt's ideas of the relationship of climatic and vegetation zones, or of the air temperature relations portrayed as isotherms.

However, knowledge gathered through instrument readings and knowledge gathered through sense impressions were not so distinct to the Enlightenment as they have sometimes appeared to be to us. Much use of instruments was designed precisely to return man to an Adamic state of sense perceptions perfect in their immediacy and accuracy. Instruments were thought of as enhancing human sense impressions rather than replacing them, were thought of as part of man's pilgrimage through the world on his way to regaining the perfect knowledge possessed by the first Adam in Eden. As the Swiss natural philosopher Jean Sennebier commented in his *L'Art d'Observer* of 1775, instruments, ". . . usually represent objects precisely; or at least they show them in the way they would appear to our senses if they were keener than they are. Instruments thus become an essential part of our senses, which they perfect, and supports to their weakness, which they reduce" (I, 3). Such a view of instruments as extensions of the human body and senses was also bound to be at odds with an interpretation of instrumental readings as being connected with the production of standardized and hence potentially universal knowledge. This was hardly a possibility in eighteenth-century conditions. Before the true mass production of instruments, it was often very difficult to calibrate readings. The susceptibility of instruments to damage and distortion in field conditions was infinite. The blurring of the distinction between human and instrument was even intensified by the end of the century, by the focus on self-experimentation in the work of scientists such as Alexander von Humboldt himself, or Johann Wilhem Ritter.

Contemporary reactions to information gathered through instruments thus do not allow us to displace persons and their locations and dislocations, from the exploration science of the Enlightenment. Instrumental readings did not substitute for the narratives of which the explorer was also the author. They also do not remove the question of the person of the explorer. Nor do they dispense us from writing the history of the acts of cognition and the categories of experience through which explorers attempted to make sense of the glittering multiplicity of the world around them. By authorship, explorers transmit these acts and categories into the European world. Thus, the way in which explorers perceive the new terrain they discover is often the founding pattern for subsequent European perceptions of the area.

In this context, let us listen again to Alexander von Humboldt, a man who had already used his own body as a Leyden jar to measure the amount of electricity generated by electric eels. Humboldt tells the same story of the convergence of the instrumental and the human in his account of his ascent of Mount Chimborazo in January 1802, a climb which in a single day made him world famous. About halfway he commented that he and his companions

> [b]egan to feel nausea and giddiness which was far more distressing than our difficulties in breathing. Blood exuded from our lips and gums, and our eyes became bloodshot. These symptoms . . . vary greatly from individual to individual according to age, experience, constitution, and redness of the skin. But in the same individual, they constitute a kind of gauge for the amount of rarefaction in the atmosphere, and the absolute height he has reached. . . . We fixed up the barometer with great care found it stood at 13 inches, 11 and 2/20 lines. The air temperature was only three degrees below freezing, but after our long stay in the tropics even this amount of cold was quite benumbing. Our boots were wet through with snow-water. . . .
>
> (Botting 1963, 154–155)

In this text, instruments and persons often substitute one for another. The individual body is "a kind of gauge," treated as valuable even in the presence of the barometer. Von Humboldt himself is highly present in the text, self-monitoring, recording instrumental readings, writing up his companions' responses and assessing their meaning. The supreme instrument present here is his witness.

The use of instrumentation is thus not the same as the history of experimentation. Emphasize the history of instrumentation, and we are unable to answer the question of what it is which makes exploration a specific form of human activity. Exploration is not merely metropolitan experimental and observational science carried out rather a long way from home. Overemphasis on instrumentation may also miss one obvious and important distinguishing feature of exploration, which is its close relationship to the definition of the new. It is this which decisively distinguishes exploration in the Enlightenment from all previous expressions in the West of the link between travel and truth, of which the myth of Perseus' search for the Gorgon is one of the most compelling examples. If the epitome of such a tradition in the Christian world is taken to be the practice of pilgrimage, then the difference becomes clear at once. The pilgrim is not traveling to find the new. He is traveling a road which will lead him but to a sacred site which is already known. An ample literature emphasized that it is the duty of the pilgrim precisely not to be curious about what he sees along the way, not to try if he can observe anything new, but to fix his thoughts exclusively on the already defined sacred goal of his journey.

The explorer in the Enlightenment is not a pilgrim. While his journey may have objectives, it cannot, by definition, have a known destination. The point of his journey is to see what has not been seen before, to pass over terrain which has not been previously traveled. It is the gaining of new knowledge in conditions of continuous transition. This distinguishes exploration from other forms of the production of the new which the Enlightenment produced. Here one thinks not only of experimental science, but also of the proliferation of taxonomic systems, whose abstraction, taken to its extreme in that of Linnaeus, was strikingly successful in acting as a diagnostic for the previously unseen. This search for the new can also be seen as a link between the experimental and the taxonomic sciences of the Enlightenment which are so often treated as separate enterprises. These few remarks clearly do not exhaust this potentially vast subject of the creation of the concepts and practices of the exemplification of the new in the Enlightenment. But this may at least be a more productive way to treat exploration in the Enlightenment, than to see it, as some postmodernists and feminists have done, as little more than an attempt to 'appropriate' the exotic natural world to European systems of understanding (for example, Pratt 1992).

Exploration knowledge is also unique in that it is produced in a situation of continuous transition. Even hostile contemporaries, such as Georges Cuvier, were in no doubt that location and dislocation were fundamental categories for understanding the nature of the knowledge so produced. For them it was a distinct category of experience, associated with a distinct sort of person. This is an important point. For quite unlike the witnessed and replicated forms of metropolitan experimental science, the authority of exploration science stubbornly came back to the person. On the sense impressions of the explorer, on his discipline, and above all on his success in welding together exactly the relation between the sensual and the moral which Enlightenment philosophers tried so hard to pull apart, was based the authority of the explorer's relationship with the new. In the end, the authority of the knowledge so produced was lodged in the authority of the person of the explorer and continued to be so at a time when other forms of scientific experience based on objectivity values were equally relentlessly pushing apart the observer and his observation. This was also at a time when the revolutions of 1776 and 1789 had equally demonstrated the possibility of the replacement of the personal authority of monarchy with the collective, representative authority of the republican systems which they founded. Against this background, far from being congruent with other forms of authority specifically developed in the Enlightenment, it

183

is possible to see exploration as representing the survival and export into exotic worlds, of far older systems of authority.

The oldest locus of authority is the human body. And it was on this that the authority of the explorer was ultimately based. Contemporary accounts return repeatedly to the theme of the authentication of the explorer's travels by the trials of his body. For example, the hair of the Berlin-born explorer Peter-Simon Pallas turned white at the age of 37 after exhausting journeys in Siberia. A fall through a treacherous ice sheet into the freezing waters of the river Ob permanently damaged his health. On returning to St. Petersburg, he found it impossible to make a permanent home anywhere, and by frequently moving house, endlessly replicated the permanent transition of the explorer's life in the field. Michel Adanson, explorer of Senegal, and author of its first flora, unflinchingly endured equivalent physical stresses and was unable to tolerate the pressure of human intercourse which greeted him on return to Paris. Thereafter a life-long recluse, his life prolonged the solitude of the field.[5] Alexander von Humboldt was permanently affected by rheumatism as a result of sleeping for months on the damp floor of the tropical rain forest. On his return to Europe, he kept his apartments in Paris and Berlin heated to tropical heights, so deeply had his metabolism internalized the demands of another continent. In another letter to his brother Wilhelm, he proudly proclaimed his physical adaptation to the tropical world into which he had chosen to journey, rather than to the Prussian plain from which he had come:

> I was made for the tropics. I have never felt so well as in the two years I have been here. I work hugely and sleep only a little. Often I make astronomical observations, hatless in the fierce sun, for five or six hours together. I have even been in towns where there has been an outbreak of yellow fever, and never had so much as a headache.
>
> (Quoted in Duviols and Minguet 1994, 15)

Men like Humboldt, Adanson, or Pallas thus did not merely bear the marks of their traveling upon the outside of their bodies; they also internalized the exotic environment mentally and metabolically. They lived in a mimesis of distance, of location in dislocation. Their bodies' suffering did not merely authenticate, in ways strikingly reminiscent of the Christian saints most of them would have repudiated as models, their claims to have traveled; they also made distant lands and oceans present to those who saw them. We all know that the body learns. And because it learns, it also teaches. What the explorer's body taught its witnesses allowed them to believe. That body was a living proof of the otherwise unseen and unseeable vastness of the world.

To say all this is not to collude with a representation of explorers as dominating, all-seeing masculine figures, which was once the darling of Victorian mythology, and is now subject to the repeated attack of the apparently equally vigorous mythologies of postmodernism. Far from it. Enlightenment explorers often manifest in their body, not domination but vulnerability. Of that vulnerability, the tattoos acquired by the sailors who accompanied James Cook into the Pacific were only the most decorative and benign reminders. Such vulnerability was usually a far more risky affair, more like walking on the thin ice of a Siberian river. But without that vulnerability the explorer could not manifest in his own person the moral economy which made his reporting acceptable as authentic knowledge. Such vulnerability, as we learned from the story of Perseus, is one of the places where action may be interrupted by reflection. It is the other side of the very love of ambitious risk that drove men onto the ice in the first place. This was acknowledged even by the taciturn James Cook, who described himself during a time of crisis in the Antarctic ice-edge search of January 1774 as a person who "had Ambition to reach furthest south . . . further than anyone had gone before . . . as far as it was possible for man to go" (Beaglehole 1969, II, 323). It was in this dance of ambition and vulnerability that the Enlightenment explorer lived. It was from this dance that knowledge of the new, such as the new continent of Antarctica, was created.

We learn from explorers that cognition has a history and politics inseparable from location and dislocation. And that the cognition of distant, new, and unseen things gains authority and reality from its incorporation in the bodies of particular men. This incorporation is not reducible to the many other ways in which exploration and explorers sponsored representations of the previously unseen. This is an important point to make not just in relation to our understanding of eighteenth-century explorers, but also to our understanding of ourselves. Today, we face very great problems about how we envision and represent the world. We seem to live in a world so awash with images of itself that the realities which these images relate to may seem less important than the games which these images of forever unwitnessed realities play among themselves. We worry about what Foucault called the "poverty of the real," and we do not replace it (how could we?) with anything which satisfies us more. At the same time, we increasingly denigrate knowledge based on direct sense impressions. We have not escaped from the dilemmas which haunted exploration knowledge in the Enlightenment. Principally, we have not yet come to terms with the cognitive and moral issues involved in making dislocated images of in fact located objects (Miller 1995; zur Lippe 1997). But in learning how to reconstruct the acts of cognition through which explorers came to know the world, in their obstinate reconciliation of the senses and the person, we also reconstruct the possibility of our own empathy with that world, and in doing so become more convinced of its, and our, own reality. The head of Medusa does not have to be our weapon.

Notes

The subject of this paper was suggested by Lorraine Daston, whom I have also to thank for her comments on several versions. I also gratefully acknowledge the helpful responses of audiences at Harvard University, the Massachusetts Institute of Technology, the University of Edinburgh, the University of Cambridge, and the Max-Planck-Institut für Wissenschaftsgeschichte, Berlin. This paper was also fortunate enough to attract the commentaries of Irina Podgorny and Wolfgang Schaeffner at the Berlin Sommerakademie of 1995. Michael Bravo and Graham Burnett made essential contributions over a long period to this paper, as did discussions with Rudolf zur Lippe. I thank Stephan Müller-Wille for illuminating conversations on Linnaeus. I take this pleasant opportunity to thank the Max-Planck-Institut for its hospitable shelter in 1995–1996, which sustained the evolution of this paper and of much else besides.

Notes

1 For another modern reinterpretation, see Klaus Heinrich, *Floß der Medusa: Drei Studien zur Faszinationsgeschichte mit mehreren Beilagen und einem Anhang* (Basel & Frankfurt am Main, Stroemfeld, 1995). A portrait commissioned by Frederick II of Prussia from the artist Christian Bernhard Rode (1725–1797) represents the monarch as Perseus about to strike the fatal blow under the direction of Athena; see Michaelis (1989, 95–96). The portrait appeared as an engraving in 1759 and was exhibited, reworked as an oil painting, in 1789.

2 A contrary argument appears by implication in Marie-Noëlle Bourguet (1997).

3 *Briefe Alexander's von Humboldt an seiner Bruder Wilhelm, herausgegeben von der Familie von Humboldt in Ottmachau* (Stuttgart, 1880), 10–18; quoted passages 11, 12–13.

4 "Analyse d'un ouvrage, de M. Humboldt, intitulé Tableaux de la nature ou considérations sur les deserts, sur la physionomie des végétaux, et sur les cataractes de l'Orenoque" (Library of the Institut de France, Paris, Fonds Cuvier, MS 3159). This passage is discussed in Outram (1984, 62–63), and, differently, in Outram (1996). An opposing interpretation is offered by Dettelbach (1993). I would like to thank Michael Dettelbach for many discussions.

5 For example, Georges Cuvier, *Eloges Historiques, précédés de l'éloge de l'auteur par M. Flourens* (Paris, 1843), 234, 231, 238, 250, 136, 148. The eulogy of Adanson was written in 1807 and that of Pallas in 1811. For the making of representative individual bodies in this period, see, in more general context, Outram (1989).

References

Beaglehole, James C., ed. 1969. *The Journals of Captain Cook, 1772–1775.* Cambridge: Cambridge University Press. 3 volumes.

Botting, Douglas. 1963. *Humboldt and the Cosmos.* London: Sphere Books.

Bourguet, Marie-Noëlle. 1997. "La collecte du monde: voyage et histoire naturelle, (fin XVIIème siècle-début XIXème siècle)," in Claude Blankaert et al., eds., *Le Muséum au premier siècle de son histoire.* Paris: Editions du Muséum National d'Histoire Naturelle: 163–196.

Bravo, Michael T. 1992. "Science and Discovery in the British Search for a North-West Passage, 1815–1825." University of Cambridge Ph.D. thesis.

Cuvier, Georges. 1843. *Eloges Historiques, précédés de l'éloge de l'auteur par M. Flourens.* Paris: Paul Ducrocq.

Daston, Lorraine. 1988. *Classical Probability in the Enlightenment*. Princeton: Princeton University Press.

Daston, Lorraine, and Katherine Park. 1998. *Wonders and the Order of Nature, 1150–1750*. New York: Zone Books.

Dear, Peter. 1987. "Jesuit Mathematical Science and the Reconstruction of Experience in the Early Seventeenth Century." *Studies in the History and Philosophy of Science* 18: 133–175.

Dettelbach, Michael. 1993. "Romanticism and Administration: Mining, Galvanism and Oversight in Alexander von Humboldt's Global Physics." University of Cambridge Ph.D. thesis.

Duviols, J. P., and C. Minguet. 1994. *Humboldt: savant-citoyen du monde*. Paris: François Maspero.

Heinrich, Klaus. 1995. *Floß der Medusa: Drei Studien zur Faszinationsgeschichte mit mehreren Beilagen und einem Anhang*. Basel and Frankfurt am Main: Stroemfeld.

Licoppe, Christian. 1996. *La formation de la practique scientifique: le discours de l'expérience en France et en Angleterre, 1630–1820*. Paris: Editions la Découverte.

Lippe, Rudolf zur. 1997. *Neue Betrachtung der Wirklichkeit: Wahnsystem Realität*. Hamburg: Europäische Verlagsanstalt.

Michaelis, Rainer. 1989. *Deutsche Gemälde, 14.–18. Jahrhundert: Staatliche Museen zu Berlin, Gemäldegalerie*. Berlin: Staatliche Museen zu Berlin.

Miller, J. Hillis. 1995. *Topographies*. Stanford: Stanford University Press.

Outram, Dorinda. 1984. *Georges Cuvier: Vocation, Science and Authority in Post-Revolutionary France*. Manchester: Manchester University Press.

———. 1989. *The Body and the French Revolution: Sex, Class and Political Culture*. New Haven and London: Yale University Press.

———. 1996. "New Spaces in Natural History," in Nicholas Jardine, James Secord, and Emma Spary, eds., *Cultures of Natural History*. Cambridge: Cambridge University Press, 249–265.

Pratt, Mary Louise. 1992. *Imperial Eyes: Travel Writing and Transculturation*. London and New York: Routledge.

Schaffer, Simon. 1988. "Astronomers Mark Time: Discipline and the Personal Equation." *Science in Context* 2: 115–145.

Shapin, Steven. 1988. "Robert Boyle and Mathematics: Reality, Representation and Experimental Practice." *Science in Context* 2: 25–58.

———. 1994. *A Social History of Truth: Civility and Science in Seventeenth-Century England*. Chicago: University of Chicago Press.

Sorrenson, Richard. 1996. "The Ship as a Scientific Instrument in the Eighteenth Century." *Osiris* 11: 221–236.

10

THE ENLIGHTENMENT OUR
CONTEMPORARY

From: William Clark, Jan Golinski and Simon Schaffer (eds.), *The Sciences in Enlightened Europe* (Chicago and London: Chicago University Press, 1999) 32–40. © 1999 by The University of Chicago.

The Enlightenment has suffered a particular fate and a particular career. No other historical period has been defined with such intensity in relation to our own. No other has so insistently been viewed as the latency period of the twentieth century. Thus no other has been so heavily defined, since 1945 in particular, not as itself but as a way of elucidating the identity of the twentieth century. The Enlightenment has become our contemporary.

It is the purpose of this chapter to examine how this view has been fashioned, and with what consequences for our understanding of the Enlightenment, and of the place of science within it. It goes without saying that such a formulation of the Enlightenment also raises fundamental questions about our understanding of what it is that we do when we use one historical epoch to comment implicitly or explicitly upon another, in this case on what we are pleased to call, somewhat possessively, "our own."

The argument that in the Enlightenment can be found the origin of our own times received certainly its most powerful and enduring statement in a work now more than fifty years old. Max Horkheimer and Theodor Adorno, in their 1944 *Dialektik der Aufklärung*, or *Dialectic of Enlightenment*, argued that the Enlightenment had carried within it forms of thought and attitudes toward nature that were the necessary forerunners of the technological horrors of the twentieth century, the Holocaust in particular.[1] In defining the Enlightenment as our contemporary in this way, Horkheimer and Adorno also defined science and technology as fundamental cultural systems that cannot be readily distinguished from the power structures of society. That is, what has defined our times as latent in the Enlightenment is the direct consequence of men's successful attempts to gain sovereignty

---

1 Horkheimer and Adorno, *Dialektik der Aufklärung* (1944); English translation, *Dialectic of Enlightenment* (1973). All references and quotations are from the 1973 translation and are cited in the text by page. Some early versions of the ideas of the *Dialectic* are to be found in Horkheimer, "End of Reason."

    DOI: 10.4324/9781003038085-11

over nature, which is no longer seen as the expression of a divine order and intention lying beyond and outside itself, but as reducible to a series of mathematical expressions. On this argument depend the other definitions of the Enlightenment given in their work, as "the destruction of gods and qualities," as "bourgeois society," as "ruled by equivalence and universalism," as "totalitarian," as "mythic fear turned radical," as a "nominalist movement," or as "wholesale deception of the masses."[2]

The situation in which their book appeared makes their interpretation entirely comprehensible. In 1944, both authors were refugees in America from Nazi persecution. They had started to confront the enormity of the Holocaust, realizing that an organized mass murder on such a scale would have been impossible to carry out without technology, and without an attitude that reduced thought to a calculus of means and ends, undercutting any independent moral ground for resistance. Insofar as the Enlightenment can be said to have laid the foundations for the development of both that technology and that attitude, it can be called "totalitarian." Both authors were also members of the Frankfurt school of social philosophers. Beginning before the Second World War, this group had aimed to use a revived Marxism to criticize industrialization and the growth of the bureaucratic state, both of which increasingly defined people as objects, and culture as a pure commodity. People and objects, as well as nature itself, were judged solely in terms of their exchange value, and not by any intrinsic worth or quality. The basis of this society was a technology that premised the complete domination of nature by man. The events of 1939–45, during which the Frankfurt school was forced to find refuge in California and New York, the onset of the Cold War, and the consequent division of Germany and Europe, caused some members of the group to abandon their earlier commitment to the revival of Marxism. Horkheimer and Adorno in particular drew back from former supporters, such as the philosopher Ernst Bloch, who remained committed to Marxist solutions. But this separation did not alter the group's determination to find a form of philosophy, a "critical theory," that would change the self-conscious actions of human beings and thus help to bring about social change. This commitment deepened with the return of Horkheimer and Adorno to Germany and their refoundation of the Frankfurt school. Because of their desire to assign meaning to the present, and working as they did at the intersection of fact and value, Horkheimer and Adorno made little distinction between philosophy and history.

The *Dialectic of Enlightenment* opened this program of engagement and criticism. The authors asked "why mankind, instead of entering into a truly human condition, is sinking into a new kind of barbarism" (xl). The answer to this question was a paradox: "[T]he Enlightenment has always aimed at liberating men from fear and establishing their sovereignty. Yet the fully enlightened earth radiates disaster triumphant. The program of the Enlightenment was the disenchantment

---

2 All these definitions appear in *Dialectic,* chap. 1.

of the world: the dissolution of myths and the substitution of knowledge for fancy" (3). Man, Horkheimer and Adorno argued, had gained sovereignty over nature and then over his fellows by "rationality," exemplified by the calculation of ends and means demanded by technology. This triumph of rationality involved a refusal any longer to see nature as the location of mysterious and inexplicable powers and forces: "Technology . . . does not work by concepts and images, by the fortunate insight, but refers to method, the exploitation of others' work and capital. What men want to learn from nature is how to use it in order wholly to dominate it and other men. . . . On the road to modern science, men renounce any claim to meaning" (3–5).

These insights link the *Dialectic* to ideas now commonplace in environmental thinking. They also engage with Enlightenment assumptions that once released from superstition and fear, once "disenchanted," humans could find purely rational solutions to problems, solutions therefore acceptable to all. Rationality would bring with it the end of politics. But of course, such agreement is never to be found, even in the calculation of means and ends. Since Enlightenment denies the validity of means, such as tradition or revelation, in solving problems, it is in the end impossible to resolve such conflicts without the use of force. This is why Enlightenment is called "totalitarian": it admits the validity of no other value system beside itself. It was not difficult to develop this point and allege that the Enlightenment had left no legacy that could be used to resist the general use of political terror, let alone the technologically ensured terror and mass death of the Holocaust.

The Enlightenment project of the "disenchantment of the world" made it difficult to focus aspirations on "the regions beyond." Rather, enlightened thought "transferred them as criteria to human organisation" (87). This transferral is what the *Dialectic* famously called "the administered life," a rational organization of men, nature, and knowledge itself for the achievement of objectives not oriented onto otherworldly or transcendent values. Rationality, defined as the calculation of ends and means, knew nothing outside itself. In support of this view, Horkheimer and Adorno quoted Immanuel Kant's dictum that "[r]eason has . . . for its object only the understanding and its purposive employment" (87).[3]

The Enlightenment idea of rationality also had important cultural consequences. It turned knowledge into a commodity. Knowledge could become part of systems of supply and demand, exchange and use, and its value could be calculated in these terms. Knowledge therefore became disconnected from what was beyond the economic sphere, that is, from wisdom and from ethics. Knowledge thus became a means to an end, and existed purely as an exchange value "without permeating the individuals who possessed it" (197). Relationships with knowledge became possessive rather than mimetic, thus losing their capacity to act critically

---

3 The quotation is from Immanuel Kant, *Kritik der reinen Vernunft* (1781).

on society. Later, in his 1970 *Aesthetic Theory*, Adorno was to argue that it is only aesthetic life that still, post-Enlightenment, allows critical moments to occur, because aesthetics contains the possibility of mimesis, of creating reminders of the transcendent world.[4]

Such arguments gave the *Dialectic* iconic status for the revolutionary and feminist movements of the 1960s. Along with works such as Herbert Marcuse's *One Dimensional Man*, it seemed to offer an explanation for the evident inability of Western society to produce an independent ground for moral action: that society had become its own object of veneration.[5] Nor was the Frankfurt school alone in raising such concerns. Simone Weil, in her posthumously published *L'enracinement* (1949), made the same case for different reasons, as did Martin Heidegger's *Holzwege* (1950).[6] The rejection of science and the values it seemed to foster took place all over Europe, and it is possible to see the crisis of the late 1960s partly as the political culmination of an ideological rejection of science, its associated value systems, and its social impact, which had been growing ever since 1945.

The arguments of the Frankfurt school in general, and of the *Dialectic* in particular, were also important because of their impact upon the rise of specifically feminist philosophies, politically and epistemologically critical, which coincided with the revolts of the 1960s, and which also strongly attacked science. Particularly important was the thesis of the *Dialectic* that the rise of experimental science since Francis Bacon was based on the search for dominance over, and possession of, a natural world essentially viewed as female. While gender is far from being a major concern of the *Dialectic*, the long shadow cast by such arguments is apparent, for example, in Evelyn Fox Keller's influential *Reflections on Gender and Science* (1985). Keller examines the gender bias in the language used by the "fathers" of modern science, which is seen as originating with Bacon.

To point out the impact of the *Dialectic* is not, of course, to imply an argument for its originality. Criticism of Enlightenment thought began with the Enlightenment itself. The famous 1784 competition announced by the *Berlinische Monatschrift* for the best answer to the question "What is Enlightenment?" revealed the unease of contemporaries with the defining concept of their time. The now best known answer to this question, that of Kant, emphasized the idea that Enlightenment was not a completed process, but one still in progress. Kant's essay opened with his well-known definition of Enlightenment as "man's liberation from his self-incurred immaturity," by which he meant dependence on external authorities rather than the use of man's reason. But he also argued that reason should be used in an unrestricted way only in the private realm. In the public realm, the unrestricted use of reason was dangerous, and led to the subversion of

---

4 Adorno, *Ästhetische Theorie*.
5 Marcuse, *Eindimensionale Mensch*.
6 Heidegger's "Question Concerning Technology" is also apposite here.

the very monarchical authority that made at all possible the security of the private sphere. So Kant's definition of Enlightenment had a double legacy for the future. It defined Enlightenment as a process, not as completed stage of human understanding. And it also pointed out central contradictions in the very idea of Enlightenment. Most important, Kant's remarks about the different ways in which critical reason should or should not operate in the public and private spheres meant that it was difficult to use the notion of Enlightenment to ground any moral theory. Would such a theory apply to the use of reason in the public sphere or the private? Would it be grounded in the public or the private roles of the individual person? If science was part of critical reason, how could it be conducted in the public realm? If it was merely practical reason, the calculation of ends and means, then it was forever divorced from the sphere of value.[7]

To the problems inherent in Kant's definition of Enlightenment, Hegel added others. In the *Phänomenologie des Geistes*, he interpreted the Enlightenment as a perverted attempt to complete the movement of the Reformation toward the emancipation of reason, which substituted for the worship of the divine the worship of human reason as the foundation of all value. In former times, Hegel argued, men

> had a heaven adorned with a vast wealth of thoughts and imagery. The meaning of all that is, hung on the thread of light by which it was linked to that of heaven. Instead of dwelling in this world, presence, men looked beyond it, following the thread to an otherworldly presence, so to speak. The eye of the spirit had to be forcibly turned and held fast to the things of this world, and it has taken a long time before the lucidity which only heavenly beings used to have could penetrate the dullness and confusion in which the sense of worldly things was enveloped, and so make attention to the here and now as such, attention to what has been called "experience," an interesting and valid enterprise. Now, we seem to heed just the opposite: sense is so fast rooted in earthly things that it requires just as much force to raise it. . . . The Spirit shows itself so impoverished that, like a wanderer in the desert craving for a mouthful of water, it seems to crave for its refreshment only the bare feelings of the divine in general.[8]

In the Enlightenment, human reason asserted the value of human reason and, without an external reference point such as that provided by the transcendent or otherworldly, was unable to provide an independent ground for morals. These arguments were important influences on the *Dialectic*.

A third major influence, this time from the twentieth century, must also be mentioned. It is to Max Weber that Horkheimer and Adorno were indebted for the description of the post-Enlightenment period as the "disenchanted world," a

7 Kant, *Political Writings*, 54–60; Baehr, *Was ist Aufklärung?;* and Hinske, *Was ist Aufklärung?*
8 Hegel, *Phenomenology of Spirit*, 5.

world from which "gods and qualities" had been banished, and in which science had joined with bureaucracy to express all qualities as interchangeable mathematical expressions. It was also to Weber that the authors were indebted for their concern with objectivity as the cultural value of science. Weber had pointed in 1904 to objectivity as necessarily and rightly the central value of the social sciences, as well as of the natural sciences. For him objectivity implied a willed distance between the opinions and personality of the observer of society or nature and the object of his observations. Weber does not argue that this definition of objectivity is necessarily separate from the sphere of social values, which alone can give meaning and value to the results of objective observation. But he gives little consideration to how meaning and value might actually be constructed. Because of this, he provided no help with the issue of how an inquiry based on objectivity might seek to ground its value outside itself.[9] Weber's discussion of objectivity thus only reinforced Horkheimer and Adorno's argument about the centrality of science to the restriction of the sphere of value in the post-Enlightenment world. This restriction, they argued, had in turn made possible the emergence of a purely instrumental form of reason, which saw knowledge as merely a tool in the calculation of means and ends, not as a means of arriving at value. These were views that enrolled their work in the quasi-Romantic rejection of positivistic science of the 1960s.

Horkheimer and Adorno thus saw the Enlightenment negatively. Enlightenment thought, particularly Enlightenment science, was the precondition for twentieth-century modernity, defined as the domination of technology and science, the restriction of the value sphere, and the consequent facilitation of technological mass murder. It is therefore interesting that their leading pupil, Jürgen Habermas, was to provide a far more positive characterization of the Enlightenment, equally influential in the historiography of science in this period. For Habermas, just as for his former teachers, the Enlightenment lay at the origins of modernity. But whereas for Horkheimer and Adorno, the central consequence of the Enlightenment was totalitarian mass murder, for Habermas, it lay in the creation of the public sphere as a location for critical reason. Ironically, however, Habermas developed this view in response to controversies over the Holocaust and the Nazi dictatorship. The German historian Ernst Nolte had argued that the concentration camp system, while undoubtedly criminal, was not unique in history, but had arisen in a reaction to the perceived threat of Bolshevism in the 1930s and 1940s. This attempt to portray Auschwitz as part of "normal history" aroused violent controversy, and provoked Habermas's intervention, culminating in the arguments of his *Structural Transformation of the Public Sphere* (1962).

Habermas emphasized Kant's insight that far from being closed, the Enlightenment even by the late twentieth century had still to be brought to completion. In

9 Weber, "Die 'Objektivität' sozialwissenschaftlicher und sozialpolitischer Erkenntnis."

his view, at least in Britain, the Netherlands, and France, the eighteenth century had seen the creation of a new "public realm" where critical public opinion was created. For all his divergences from the *Dialectic*, Habermas still agreed that the Enlightenment was characteristically bourgeois. And whereas Kant's essay had emphasized the limitations on the public sphere in states such as his native Prussia, Habermas argued for its efflorescence further to the west, a result of middle-class control and consumption of the flow of cultural materials, which became so important as commodities in this period. While Habermas's closing attack on modern mass culture differs little from that of Horkheimer and Adorno, he does nevertheless see the bourgeois public realm as a place of liberation, where critical reason can question traditional social forces. Habermas's public realm is a space where men can escape their role as subjects. There is no need to emphasize how important this view might be to a German historiography traditionally overshadowed by the creation of the state, and overwhelmed by the problem of the failure of German civil society and liberal humanist values to resist the rise of Nazism.[10]

In spite of the long gap between its first publication and its translation into English, and in spite of the fact that it deals explicitly only with France, Britain, and the Netherlands, Habermas's work has been influential for the history of science in Enlightenment Europe. In the historiography dependent on his work, science is often viewed as part of this revolution of media and consumption, part of the commodified flow of information that characterized the Enlightenment. This released historians from the dark view of science and technology put forward by Horkheimer and Adorno. It also released them from the internalist agendas of a previous generation of historians, and allowed them at least to pose the question of the relation between the history of science and the newly fashionable field of cultural history. However, the way in which historians such as Robert Darnton or Roy Porter have used Habermas's work has often made it difficult to assign a role to scientific ideas as such. Whereas Horkheimer and Adorno were sure that science and technology, for good or for ill, had supplied the central cultural structure of the Enlightenment, Habermas gave no specific place to science in the public realm. Historians who follow Habermas can find no way from within that theory to explain why, or whether, science could be assigned this centrality. The description of the Enlightenment as so strongly linked to the creation of middle-class self-identity through a relationship with print media also begs many questions. In the hands of the unwary, it can provide a license for seeing the Enlightenment as an innocent creation, the product of reading, writing, and conversation, of amateur science and genteel amusements: something that happened in the free time of the majority of its consumers, on their holidays from the realms of power and authority. The merit of Horkheimer and Adorno's approach is to make it impossible to divest consideration of the Enlightenment from the consideration of power.

10 Habermas, *Structural Transformation of the Public Sphere;* Eley, "Nazism, Politics, and the Image of the Past."

It is also important to note that Habermas's approach provides no solution to the problem of the relationship between the history of science and cultural history. It thus becomes difficult to justify the preference for portraying the Enlightenment as in some sense the origin of modern science.

What claim do we really make in such portrayals? Are we merely concurring with the insights of Horkheimer and Adorno into the relationship between the Enlightenment, modernity, and science, insights that seem to provide very little way of understanding science critically but positively, and thus enabling us to emerge from the current aggressive polarities of the science wars debates. How can we construct a history of science that is different?

A first step might be to ask ourselves what it is that we do when we describe another historical period as our own origin, in this case, when we describe the twentieth century as latent in the eighteenth. When we do this, we are identifying ourselves as contemporary to a certain time we think of as "ours," and we are simultaneously saying that the Enlightenment, connected to that time, is therefore also, to greater or less extent, "ours." Such thinking is highly possessive. It is also full of notions of identity, an idea the *Dialectic* repeatedly criticizes. By clinging too hard to such ideas as the justification of our interest in the Enlightenment, we run the risk of approaching it simply as a mirror to ourselves, of sacrificing its specificity to the need to find projections of ourselves in the past. In this sense, and with great irony, such assertions of contemporaneity have in fact a damaging effect on the possibility of writing the history of science as part of a critical cultural history, because they remove the possibility of the formation of external points of reference to validate that critical viewpoint.

The Enlightenment chose as its own projection field, not a historical era relatively close to it in time, but the societies of classical Greece and Rome. It was here that the era found its political and literary models. Other widely used projection spaces were described in the travel literature and utopian journeys of which the Enlightenment produced such great numbers. In other words, for the Enlightenment, very differently from our own period, origins were placed precisely through distance, not through some form of possessive identity. This is a perception we should consider before we designate any historical period as "ours" – and hence ourselves as "its"?

Primary sources

Hegel, G. W. F. *The Phenomenology of Spirit*. Translated by A. V. Miller. Oxford: Clarendon Press, 1977. First German edition published 1807.

Kant, Immanuel. *Kant's Political Writings*. Edited by Hans Reiss. Cambridge: Cambridge University Press, 1970.

11

HEAVENLY BODIES
AND LOGICAL MINDS[1]

John Banville's astronomical novels

From: David Attis and Charles Mollan (eds.), *Science and Irish Culture Volume I: Why the History of Science Matters in Ireland* (Dublin: The Royal Dublin Society, 2004; ISBN 0860270475), 19–25.

John Banville's astronomers, Kepler and Copernicus, shiver alone in icy rooms in high towers or in the back rooms of small, crowded family homes, undertake strange journeys across the frozen, empty central European landscape, endure ridicule from Emperors and dwarves, exercise solitary authority, make overtures to those around them that end in mutual incomprehension, destruction or a precarious warmth. Alone, they live uncomfortably and ineluctably with the elusive patterns inscribed by the stars. Banville's heroes in fact go through all of the torments involved with the full realisation and implementation of a scientific vocation. Apart from the title study of Ronan Sheehan's *The Boy with the Injured Eye*,[2] Banville is unique in modern Irish literature in his insistent return to this theme and in the intensity of his portrayal. The astronomical novels, along with the ironic commentaries on them formed by the 1982 *The Newton Letter*, are incontestably the high peak of Banville's achievement in the intensity of their focus, their tight intellectual control and their highly charged language.[3]

Unlike the earlier *Birchwood* (1973), with its themes of entangled relations and decaying great house,[4] they are not identifiably 'Irish' novels. And here emerges the first major problem. All novels contain implicit discussions of systems of values. They create images of existence that seemingly possess reality and thus invite the reader to contrast the values manifested in the novel with the values of the reader's own 'real' experience. To write about the astronomers Kepler and Copernicus in their lonely obsession with pinpricks of light that pierce the

---

1 Originally published as Dorinda Outram, "Heavenly Bodies and Logical Minds," *Graph* (1988) 4: 9–11.
2 Ronan Sheehan, *The Boy with the Injured Eye* (Dingle: Brandon, 1983).
3 John Banville, *Dr Copernicus: A Novel* (London: Secker and Warburg, 1976); *Kepler: A Novel* (London: Secker and Warburg, 1981); *The Newton Letter: An Interlude* (London: Secker and Warburg, 1982).
4 John Banville, *Birchwood* (London: Secker and Warburg, 1973).

DOI: 10.4324/9781003038085-12

darkness is to hold up for inspection values of personal autonomy, of the willing conflation of the individual subjectivity with the construction of knowledge of the most 'objective' and yet most challenging kind – that concerned with the movement of distant planets.

Few constellations of values could be more distant from the 'Irishness' of the themes of *Birchwood*. Banville seems to be coming to grips with this gap in the more recent 'scientific' novel, *The Newton Letter*. Much shorter than *Kepler* and especially than the longer *Copernicus*, *The Newton Letter* does not deal directly with the life of Newton himself but frames moments of the great natural philosopher, often troubled or verging on madness, by the narrative, which concerns the efforts of the hero, a historian, to complete a study of Newton. Needing solitude and autonomy, he rents a cottage on the estate of a decaying 'great house' near Ferns and promptly falls into increasingly complex and disastrous relations with his landlords. As his personal life collapses into chaos, invasion and confusion, so work on the Newton book slows down and eventually ceases altogether, only to be completed far away in a midwestern university.

Newton, in other words, seems to collapse the values he embodies as yet another lonely, natural philosopher prepared to conflate self with the patterns of the distant planets, powerless against the relentless, intertwining subjectivity of the narrator and of the three occupants of the Big House. But at another level, the incompletion of the Newton book and the confusion of the relationships between the occupants of the Big House and the historian-narrator in fact stem from the same source: an incomplete capacity on the part of the hero to see the world about him for what it really is; to be as clear-sighted as his astronomer-subject. The narrator's relations with his landlords Edward, Ottilie and Charlotte are falsified from the very beginning by his insistence on viewing them and their situation through the murky lens of a very 'Irish' historical and literary stereotype:

> I had them spotted for patricians from the start. The big house, Edward's tweeds, Charlotte's fineboned slender grace that the dowdiest of clothes could not mark, even Ottilie's awkwardness, all this seemed the unmistakable stamp of their class, Protestants of course, landed, the land gone now to gombeen men and compulsory purchase, the family fortune wasted by tax, death duties, inflation. But how bravely, how beautifully they bore the losses![5]

From this initial and total misreading stem many of the complications and disappointments of the rest of the novel. And unlike the heroes and heroines of many other novels of misapprehension, like *Northanger Abbey*, for example, the narrator of *The Newton Letter* makes no gradual progress into maturity, enlightenment and a correct view of the relationships that surround him. His final comprehension

5 Banville, *The Newton Letter*.

is sudden, overwhelming, and entails not his reconciliation with Edward, Ottilie and Charlotte but his abrupt departure from them and their world. In the end, the value-system of the book enhances, more subtly than do *Kepler* or *Copernicus*, the value of the clear gaze, the value of the refusal to entrust one's subjectivity to a falsified world of relationships.

It is in this sense that Banville's novels, in their concern with the limits and control of subjectivity, with the problem of how human beings can produce rationality out of their inherent subjectivity, are projects concerned with Enlightenment values, of which scientific rationality and the whole imagery of light, vision, gaze, seeing and ultimate enlightenment in the full sense, are essential constituents. These are concerns he shares with Ronan Sheehan, whose 1983 study of the 18th-century Limerick ophthalmologist Sylvester O'Halloran is a continuous examination of the real meaning of what is involved in clearly seeing the world around one. Towards the end of his life, O'Halloran reflects:

> Was his work as a surgeon really so valuable or significant? What was the point of refining methods of helping people to see when things were so miserable that they wanted to close their eyes? In removing a cataract all one achieved was the readmittance of light which reflected the actual world, the outer world, lost and in darkness. Would it not be more valuable to do as (the Gaelic poets) Seán Clárach and Seán O Tuama did, to provide an alternative inner world, brilliant with light?[6]

It comes as no surprise that themes such as science, the Enlightenment, and the kind of enlightenment that comes from the management of personal subjectivity into personal autonomy should begin to appear in Irish writing after many years of debate in European critical theory on the nature and legacy of Enlightenment values, and at a time when movements loosely jumbled together as New Right place increasing emphasis on the achievement of radical personal autonomy for individuals.

In the Irish case, more specifically, it is also becoming clear in the work of many critical commentators that the absence of Enlightenment values of rationality and progress and the absence in the culture of support for ideals of radical personal autonomy are two of the major factors accounting for Ireland's failure to modernise in the 20th century.

Banville's novels also leave us confronted, however, with equally pressing questions about the insertion of science itself into Irish culture. To begin with, science has risen to its present dominance both as value-system and as a central factor in the state and society of the 20th century as much by a series of propaganda plays to control culture, as by actual 'delivery' of 'facts' or 'discoveries'. Central to these controlling ploys was the claim that science itself was a free

6 Sheehan, *The Boy with an Injured Eye.*

space, value-free and passionless in the midst of a world riven by chaotic social and personal contests and characterised by the coercive manipulation of one individual by another.

Science was the neutral ground, where exploration of nature was most effectively conducted by the individuals who had achieved the utmost personal authority removed from the necessity of a negotiation with their fellows. The pure in heart were the pure in eye, able to 'see' nature as it really is. This myth has cast a long shadow leading directly to the still continuing view held by many scientists even today that 'science' has nothing to do with 'politics', and the issue of the moral responsibility of scientists for their involvement with such highly politically charged projects as the creation of the atomic bomb or the development of artificial means of human reproduction can therefore be conveniently pushed to one side.

In a more positive light, however, the myth of the 'autonomous' person as the ideal, unmoved interrogator of nature, formed part of a continuous drive in Western European culture, which gathered in intensity throughout the post-Reformation period and achieved its peak in the era of the French Revolution, to create images of solitary, autonomous and thereby 'authentic' man. The solitary scientist is simply one of the many variants of that 18th-century best-selling hero, Robinson Crusoe.

With the concepts both of science and of individual autonomy, Irish culture has manifested great unease. The failure of Ireland to develop 'big science' in the 19th and 20th centuries was probably inevitable, due to the financial demands of such an undertaking and the absence in the Irish case of the pressing demands for the scientific and technological advance of modern warfare. This goes far to explain why most of the men of science working in Ireland who are still remembered today by historians of science practised in the 'cheap' sciences – like mathematics (William Rowan Hamilton, George Boole), or those which, like astronomy, were cheap to equip in the 19th century (Hamilton again). Once an era opened after 1914 where scientific advance became a matter of even more complex and costly equipment, practically necessitating state intervention on a large scale, Irish science began to look very old-style indeed. No heroes of the solitary search for natural truths, no Curies or Einsteins, no Niels Bohrs could emerge.

There were also, of course, other reasons why scientific heroes are not to the fore in Ireland. The emergence and eventual triumph of the nationalist tradition led to a continuous reworking of the national pantheon. Because of the influence of the Catholic Church on that tradition, natural knowledge (and therefore its producers) was viewed with deep suspicion, as potentially leading to materialism and atheism, the worship of the creation, rather than the Creator.

The relative indifference of the Catholic tradition as such to the natural theological enterprise of asserting the very existence of God through analogy with the nature of His creation also meant that science could find little means of accommodation with what, by the 20th century, had become the national consensus. Natural science quickly settled down to an eternal repetition of 19th-century descriptive practices based on the field rather than laboratory sciences. Scientific popularisers, such as the Protestant Robert Lloyd Praeger, whose works from the 1920s and 1930s are still reprinted, were careful to attempt to legitimise the field sciences by the readiness with which they could be combined with antiquarian and linguistic or folkloric enquiry into the 'national heritage'.

But that 'national heritage' was already one conceived in terms of fundamental binary pairs of opposites: as Catholic and thereby anti-Protestant; as Gaelic Celtic and not English; as republican, not monarchist; as conservative and therefore not modernising; as culturally monolithic, opposed to the pluralistic cultures of the rest of the West: and through a combination of all these factors, based on history, folklore and literature, rather than on science – validating Seán O Tuama rather than Sylvester O'Halloran.

Due to the way in which Irish national culture was constructed in conscious opposition to Enlightenment values, science came to be identified with all that Ireland, in its own nationalist tradition, was not: Protestant, modernising, foreign, and thereby associated with the Ascendancy. In more recent times, this position might seem to have altered. The pressures of European Community membership and of a chaotic economy have indeed produced much rhetoric from government, universities and industry about the need to 'modernise', to become 'technological'.

At the same time, smaller, informal but vocal groups contribute much to public awareness of the importance of the application of scientific and technological intervention to the environment. Yet important as this development has been, it has remained identified with a perceived set of values. In objective terms, Ireland fails to generate 'big science' and remains the mere location upon which science and technologies produced and financed elsewhere are practised. In intellectual terms, as well as meteorological ones, Ireland remains subject to a Windscale effect – not generating power but suffering fallout.

It is also the case that the Irish Catholic Church, still the dominant social and cultural force in the nation, has produced neither theology nor informal public pronouncements on science as such or its implications. Concentrating on one single natural object – the body and more particularly its reproductive history – the Irish Church has jettisoned the entire theological tradition that insists on the whole natural order as a demonstration of the existence and the nature of God. It possesses little ability or desire to try to integrate 'science' into the national culture.

Science in Ireland thus remains fixed in a culturally peripheral position not only by objective realities, like a lack of government funding, or by the hostility of bodies like the Church that make claims to cultural hegemony, but also because it remains locked in a colonial mindset, identified willy-nilly with the 'Protestant' moderniser rather than the 'Catholic' traditionalist.

Against this background, Irish writers who wish to treat science seriously encounter severe problems. To create a fresh literature, they have to abandon the well-worn post-colonial dichotomies. But this matter of abandonment cannot be a quiet affair, a slow drifting in the wind away to some non-Irish cultural location. It has to be done more explosively because of the enormous and increasing gap between the nature and needs of modern Ireland and its nationalist cultural heritage. In order to maintain cultural hegemony for the Catholic-nationalist tradition, many values of prime importance in the rest of the world since the period of the Enlightenment have been sacrificed; chief among these is the idea of personal autonomy as a value, and the idea of individual subjectivity as a good.

Because Banville writes novels about science, he ends by privileging those very two factors simultaneously from two different angles. First, and very obviously, to write a novel about any one individual implies the building of an entire universe, within the novel, around the subjectivity or mythology of science that focuses on the unique claims of autonomous individuals to act as the only true interpreters of nature. That such a mythology of science is just as much a fanciful production as the mythologies of Irish nationalism does not alter the fact that by imploding the fantastic representation of science into a literary tradition still very much concerned with the tradition and image of 'Ireland', Banville, intentionally or not, has set up a counter-novel, a counter-mythology, that challenges many of the most basic assumptions of the nationalist culture. And it does so even more profoundly than it could by attacking its post-colonial presuppositions explicitly.

Banville's novels *Kepler* and *Copernicus* are not simply novels; they are narratives about individuals who really existed and who are themselves the focus of a long-standing tradition of scientific biography – and hence mythology. As we have seen, it is the central feature of that mythology to place individuals valued in virtue of their autonomy at centre-stage in a historical setting, because the nature of their subjectivity enables them to emerge as successful interpreters of the natural order. In the novels, Banville shows us characters whose subjectivity is so powerful that it does in fact change the course of a history that exists not only inside this novel. As all novels contain fictional worlds with fictional histories – in a way that means that the reader is aware of their participation in the history long after he or she has closed the pages of the novel – the impact demanded on the reader is to recognise that they live in the worlds that Kepler and Copernicus have fashioned.

The novel in fact escapes from the confines of the history of Ireland and lets in other histories that have as their subject the counterpoint between complex and densely described historical situations and the tracing of the evolution of the subjectivity of the hero, Kepler or Copernicus, into the state of complete absorption and oneness with the understanding of the cosmological problems they have chosen for their own. The very contested struggles of individuals for individual direction in a hegemonic culture that are the topic of writers such as John McGahern are simply ignored in Banville's astronomical novels. There, the outside world is perceived as unhegemonic, incurably chaotic and threatening. Individuals grow to be themselves, which means in the case of Kepler and Copernicus, to make their perceptions of the natural order part of themselves, not as a struggle against an authoritarian order that privileges family, religion and tribe against individual perception.

Banville's bringing together of exterior social chaos and the subjectivity that imposes patterns on the universe is best expressed in the closing pages of *Kepler*:

> When the solution came, it came, as always, through a back door of the mind, hesitating shyly; an announcing angel dazed by the immensity of its journey. One morning in the middle of May, while Europe was buckling on its sword, he felt the wing-tip touch him, and heard the mild voice say I am here. . . . Two weeks after the formulation of that law the book was finished. He set about the printing at once, in a kind of panic, as if fire or flood, his greatest fears, or some other hobgoblin, might strike him down before he could make public his testament.[7]

Such is the weight that the subjectivity of Banville's historical astronomies actually imposes on the world, that at their most achieved form, to jettison history itself becomes possible. As Kepler reflects on himself,

> For years the World Harmony had obsessed him, a huge weight pinning him down; now he was aware of a curious feeling of tightness, of levity almost . . . that was the demon. He recognised it. He had known it before, the selfsame feeling, when, in the *Astronomia Nova*, he had blithely discarded years of work for the sake of an error of a few minutes of arc, not because he had been wrong all those years – though he had – but in order to destroy the past, the human and hopelessly defective past, and begin all over again the attempt to achieve perfection: that same heedless, euphoric sense of teetering on the brink, while the gleeful voice at his ear whispered jump.[8]

Banville's scientists, in other words, manage at the end to obtain some giddy freedom from the real 'history' that weighs down the Irish novel and prevents the ultimate resolution between subject and object, autonomous person and natural order.

Banville's scientific novels thus stand outside the mainstream of narratives on science produced on either side of St George's Channel. The British have produced narratives *about* science, from Erasmus Darwin and Mary Shelley, in the 18th century, through to H. G. Wells and Aldous Huxley in the 20th century. But only one author, Mary Shelley, made her narrative on science also a narrative on individuality. Conversely, the Irish have produced almost no narrative renderings of the scientific pursuit apart from Banville's, which gain their charge precisely because they implode yet again the question of subjectivity both into the world of big science so evident across the water and into its double, the Irish 20th century, which imagines that modernisation is possible without the acceptance of the Enlightenment and its hero, the authentic, autonomous individual.

7 Banville, *Kepler*.
8 Banville, *Kepler*.

12

NEGATING THE NATURAL

Or why historians deny Irish science

From: David Attis and Charles Mollan (eds.), *Science and Irish Culture Volume 1: Why the History of Science Matters in Ireland* (Dublin: The Royal Dublin Society, 2004; ISBN 0860270475), 27–31.

By now, it will hardly bear repetition that Irish culture is deeply involved in history-making, and involved in a way which is unusually tense, contorted, and emotional. The 'revisionist' historians, who are now, with the passage of time, turning into a new traditionalism, have, at bottom, done little to question or shift the nature of this relation. Revisionism might well be producing a better, less obsessionally Anglo-centred history: but it has not effectively questioned the central role of a certain sort of history-making. Concerned to reassess the political and social impact of Britain on Ireland, they have rarely moved squarely outside the framework of concerns established by the other, nationalist school of historians. As a result, entire areas of experience have been largely dispossessed of their history. Foremost amongst these areas is that of science, in the broadest sense the history of the human encounter with the natural world. Now an expanding and prosperous area of both academic and popular interest in America, Australasia, and in a very different tradition in continental Europe, the history of science has in Ireland only a few scattered professional practitioners. In the Republic there is no university department concerned with it. Although the Royal Irish Academy maintains a History and Philosophy of Science Committee, with a membership drawn from both sides of the border, and although there have been some recent conferences on Irish science, it is difficult to deny that the institutionalisation of history of science is weak, and its impact at the crucial level of undergraduate teaching, is minimal.

Few Irish historians have challenged this state of affairs, because to do so would be to challenge the 'deep structures' not only of Irish history, but of Irish historical scholarship and of Irish culture. We need to understand how the whole area of man's encounter with nature has been dispossessed of its history, because this dispossession reflects a critical failure both for science and for history in the Irish context. Lacking a historical dimension, science is deprived of an essential input into future policy planning, both as it affects the making of science itself, and its implementation in the social world. Conversely, history without a history of

DOI: 10.4324/9781003038085-13

science is not merely refusing the task of accounting for one of the most dynamic forces shaping the modern world; it is also refusing the task of critically evaluating its understanding of human beings as political animals. Irish history has always focused, to the detriment not only of history of science, but also of social history, and of the history of *mentalités*, on a narrowly-conceived version of the transactions of power. To a large extent it refused to examine un- or anti-political ideologies, such as that which for centuries has been at the heart of the overt ideology of scientific vocation. It has also squarely located power as a resource drawn from human relations of dominance and deference, rather than critically examining other ways in which authority may be obtained. But for Irish history to deprive itself of the history of science is also to deprive itself of some valuable tools for critical reflections on the nature of power and authority, and the capacity of man to gain that power and authority through contact with sources which, like the natural world, stand outside the human group. It is also the case that while revisionist history ignores the history of science, it will not only be writing the history of the dispossessed: it will itself be a dispossessor.

It is easy enough to assemble reasons for this failure of the Irish historical imagination. Unlike the American historical tradition, and in spite of marked regional differences in settlement patterns and prosperity, Irish historiography lacks the concepts both of the 'frontier' and of the 'garden', two crucial images of the social mediation of the natural. Even more importantly, Irish socialism has never betrayed much interest in the linkage between science's alleged gifts of objectivity and rationality and the socialists' dream of an equal and rationally planned society. It is not only that social radicalism has always been weak in Ireland, but that science does not have a long-standing input into a political tradition which would at once validate and make accessible those values, and force them into historical consideration. The absence of science from the Irish political tradition as a source of ideology points up the isolation of even Irish radical politics from the continental mainstream, where science-as-progress-reason-and-justice has been present ever since the days of Condorcet's *Sketch for the Progress of the Human Spirit*. The absence of historical comment on this state of affairs highlights the extent to which, even in the act of critical destruction, revisionist history has absorbed the pre-suppositions of the historiography and the mentality which it sets out to dismantle; such as that power is all, and authority nothing; and that there are no experiences except those which occur solely as a result of negotiations within the human group.

Given history's failure in this region, it becomes less surprising than it otherwise might be that the only contemporary Irish exploration of one of the central problems of science's history, the inner experience of a scientific vocation, and its implementation in the real world, should have come from a literary source, from John Banville's two extraordinary recent novels, *Kepler* and *Copernicus*. So peculiar is a culture which deifies history yet yields to literature the vital tasks of rendering visible heroic models of the scientific pursuit and the scientific vocation, that we must seek the origins of this situation not merely in the specific problems of Irish historiography. Even revisionist history has not yet fully realised that Irish culture, the very culture which it believes it is helping to change, has a distinctive attitude towards the natural world, an attitude which enables and encourages it to marginalise both the history of man's encounter with the natural, and the experience of nature itself. Everything, that is, except the human body.

Recent timely statements such as Dolores Dooley's UCC television lecture 'Expanding an Island Ethic', in the series *Towards a Sense of Place*, have exposed at great length the now traditional Catholic equation of moral theology with the control of bodily events – conception and its control, birth and its prevention, sexuality in all its aspects, illness and surgical intervention, death and dying – in the shape of hospital Ethical Committees. The body appears in this traditionalist account as a place where the ability to control is overwhelmingly made manifest. Such a control – until very recently – has made attitudes to the body a touchstone of social conformity. An accompanying feature of this emphasis, almost its intellectual pre-condition at the level of technical theology, is a correspondingly very small interest by the Irish Catholic Church in any form of natural theology. Absent from most current theological pronouncements are references to the natural order as evidence of Divine intention in creating the world for man to inhabit. The only element of this natural order which is routinely, even insistently, referred to, is the human body itself. In comparison with the eighteenth-century churches who saw man as an integral part of the entire hierarchy of the Great Chain of Being stretching from stones and earthworms up to the angels themselves, the ideological impoverishment is obvious. Since the eighteenth-century the Church has dramatically restricted that part of the natural world which it integrates with its moral culture. In the peculiar circumstances of Ireland, such restriction and impoverishment have inevitably become embedded in the culture as a whole. Concern with the *control* of bodily events has seemingly inevitably been accompanied by a screening out of any debate of man's control over, or active intervention in, the shaping of the natural order. That environmental issues have little or no place in debates recognised as politically crucial in Ireland today is a significant fact. Instead 'the natural' is allowed a role in political culture only in the area of individual management of the body. There is little interest in either the conservative or the liberal camps in issues, such as environmental ones, which take choices beyond the individual level, and involve the discussion of man's relation through politics with the *entire* natural order.

The peculiar circumstances of Ireland have also meant that discussion of individuals' control of their own bodies and the closely related issue of sexual morality have taken over from discussion of political morality. Such diversion of time and energy onto discussion of a tiny segment of the natural order has had an important and conservative function in Irish politics. It also forcefully demonstrates the failure of the Irish state to achieve the autonomy from discussions of individual morality which most would see as one of the hall-marks of modernity.

But if a history is to be produced which can transcend and critically examine these long-term trends in Irish culture, what would that history look like, and what

205

questions would it ask? Its first task must be to establish, contrary to many trends in 'revisionist' history, what is *specific* to the Irish encounter with the natural world. What has the exercise of a scientific vocation entailed at any given time? What did it, and what does it, mean to make a 'career' in Irish science? What sort of substantive science has been produced in Ireland? How, if at all, was enquiry into the natural world linked to the elaboration of any of the strands of the nationalist tradition? How did the pluralist religious tradition of Ireland help to have an impact on theological justifications for the study of nature? – a question which, as I have hinted before, is not without interest in accounting for the impoverishment of the modern Catholic hierarchy's confrontation with 'nature' in the shape of the human body. There is a great need too for more empirical work on Irish science and its history, a task undertaken almost alone for many years by Gordon Herries Davies of TCD [Trinity College Dublin]. We need to know, for given periods and places, the number and nature of scientific 'practitioners'; the audience for science; the public image of science; its place in government policy and funding. But empirical work is never enough. Historical enquiry is meaningless unless it can also suggest explanatory models which link problems and societies together. This is also the only way that the history of small nations can bring a conceptual purchase to bear on the history of other states. This would involve far more than the scaling-down or re-cycling of theories about the historical development of large states, an intellectual version of the way in which third-world peoples turn the detritus of the first world – Coke tins and discarded tyres – into useful, and different, artefacts. It would entail the creation of new models of scientific development, as well as the identification of gaps in the current fabric of explanation. For example, one of the most insistently touted theories of the emergence of modern science describes a Faustian compact between a highly professionalised science and a state funded 'military-industrial complex'. This model is of doubtful relevance to a history of the emergence of science in Ireland, just as it is of doubtful relevance to the emergence of science in weak, partly colonial and disunited states such as Italy. Like Ireland, Italy specialised in the production of isolated 'flash-points', often in the physical or mathematical areas. Isolated genius like the physicist Fermi in the 1920s and 1930s, the electro-magnetic pioneers Volta and Galvani in the late eighteenth-century, produced work of European resonance, which failed to leave behind either a mature research school, or a high degree of institutionalisation. Their cases are closely paralleled by the Irish, or Anglo-Irish, mathematicians George Boole and William Rowan Hamilton. The Italian comparison is important because it also shows us an alternative to the otherwise easy option of adopting the 'colonial science' models currently so fashionable in North America and Australasia. It is unclear that Ireland *was* in fact a colonial society in the same sense as were, for example, Canada or Australia. It feels closer to the 'late-emerging' nations of the European mainland, which like Italy, in spite of enjoying semi-protectorate relations with other great powers, also had strong, pluralist, conflicting indigenous cultures of their own. Such states, with strong Catholic traditions and limited national resources, face problems which are far closer to the Irish experience than are those of the vast territories of Australia and Canada. Contrast with societies like Italy can, properly employed, tell us far more about the reasons for the specific national characteristics of Ireland. For them we may have to apply models closer, for example, to Schumpeterian ideas of sudden perturbations to cultural forms and economic systems.

It is not my intention here to work out such a model in detail; but in conclusion, such an emphasis on model-finding, with the objective of explaining the relationship of Irish culture to the natural world, would have a very specific contribution to make to current political debates which almost all revolve around attitudes to the biological events of the human body. We need far more understanding of how these attitudes, which even in theological terms represent an enormous restriction over earlier models on the range of possible engagement with the natural world, have come about, have been accepted and have survived. Without such an historical understanding of attitudes to nature, we lack a crucial perspective on current struggles, and we lack knowledge too of the reception of science itself in Irish culture, surely an important element in the formulation of science policy. These are useful questions for a time when only the seizure of the *effective* social and ideological implantation of technological opportunity will prevent Ireland becoming, in all but name, a truly colonial state. All this can only happen when Irish culture in general, and Irish history-making in particular, stop saying no to the natural.

13

ENLIGHTENMENT STRUGGLES

From: C. Scott Dixon and Bean Kümin (eds.), *Interpreting Early Modern Europe* (London and New York: Routledge, 2019), 417–438. © Dorinda Outram.

Introduction

No period in western history has been as keenly debated as that of the Enlightenment. Since the eighteenth century itself, historians, philosophers, politicians, and literary scholars have tried to find its meaning, and assess its impact on the world in which they lived (Edelstein 2010). As we will see, they have often tried to use the interpretation of the Enlightenment as a way of jumping from historical fact to moral and philosophical imperative, from the realm of Is to the realm of Ought. In doing so, they have stood within longstanding debates about the uses of history, debates which run from the Classical world into our own days. It was only in the nineteenth and twentieth centuries that, for a brief time, historical knowledge was supposed to be 'objective', or 'scientific'. More recently, it has become fashionable once again for the historian to place him or herself inside the text, and turn commentary on the Enlightenment into a gloss on current trends and problems, not to mention burdening their readers with autobiography. Philosophers such as Richard Rorty, and historians such as Anthony Pagden, Jonathan Israel, and Vincenzo Ferrone, for example, have argued that the Enlightenment lies at the foundation of modern secular cosmopolitan liberalism.

Yet not all commentators have ever taken on board the Enlightenment's selfportrait as a campaign for utility, reason, tolerance and progress. As early as the 1790s, the Enlightenment was blamed by conservative commentators, such as Joseph de Maistre, for the violent excesses of the French Revolution and its attack on throne and altar, which, they alleged, had caused the collapse of the Ancien Régime. In 1947, with their *Dialectic of Enlightenment* (*Dialektik der Aufklärung*), the refugee German philosophers Max Horkheimer and Theodor Adorno, in the immediate aftermath of the Second World War and the Holocaust, argued that the Enlightenment had forged the technological, 'instrumental', ends-orientated reason which allowed the organisation of man-made mass-death and the ruthless exploitation of nature.

This list of negative interpretations could easily be expanded. By themselves, they leave us with a single question: why has the Enlightenment, easily more

DOI: 10.4324/9781003038085-14

than any other field of western history, attracted this 'interpretative overload'? Its proximity to two major crises of modernity, the French Revolution which created the modern political individual, and the Holocaust which called into question the very meaning of the terms 'reason' and 'humanity', may go part way to answering this question. For the nineteenth century, the French Revolution, allegedly caused by the Enlightenment, was the defining issue of its time; for the twentieth century, and arguably for the twenty-first, it was the Holocaust which occupied that role. In the twentieth century, Horkheimer's and Adorno's powerful argument about Enlightenment terror for at least two decades was dominant, and fed into rising anti-science, anti-colonialist, feminist, and leftwing views in history and philosophy from the 1960s onwards. *Dialektik* in hand, it was easy then to support the view that the Enlightenment's allegedly 'instrumental', value-free, ends-orientated reason had led not only to the Holocaust, but to the exploitation of nature, and exploitative attitudes towards women and non-European races. Hostile interpretation of the Enlightenment became integrated into the newest and most passionately debated political lines of the post-War and post-imperialist era in the western world. Once more, the Enlightenment bore a load of interpretation very different from that of other historical periods.

It was easy enough to find the very words of some Enlightenment thinkers to confirm this point of view, for the Enlightenment itself had not been without internal critique, which again gave hooks for multiple interpretations. The great German thinker Johann Gottfried Herder (1744–1803), for example, challenged Enlightenment meliorist views of human history, which he called:

> The general philosophical, philanthropic tone of our century, which wishes to extend our own ideal of virtue and happiness to each distant nation, to even the remotest age of history . . . It has taken words for works, Enlightenment for happiness, greater sophistication for virtues, and in this way invented the fiction of the general amelioration of the world.
>
> (Herder, quoted in Barnard 1969, 187)

Herder believed that in this way the high-minded men of the Enlightenment had justified the dominance of European culture over that of other races. As he wrote: 'The ferment of generalities which characterize our philosophy, can conceal oppressions and infringements of the freedom of men and countries, of citizens and peoples' (Ibid., 320). It was difficult, for example, to see how the Enlightenment could have set up a bulwark against the evil of African slavery. Its reiteration of belief in the dignity of man sat oddly with the absence (except in England) of organised opposition to the institution of slavery until the end of the century.

Rather than being a unified movement of thought, the Enlightenment was full of contradictions. Ideas were often not carried to their logical conclusions, and practice and theory often swung widely away from each other. Beliefs about the inferiority of women and non-white races, for example, were difficult to fit into Enlightenment generalisations about the dignity of humanity, and human rights.

This may also go far to explain the 'interpretative overload' which we have noted before. All who wished could find in the Enlightenment a hook for their own views, or fulfil their interpretative needs in the construction of a world philosophy or outlook, whether their views of the Enlightenment were positive or not.

In recent times, much more emphasis has been placed on the idea of an extra-European Enlightenment; historians have also spent much more time on constructing a gendered Enlightenment. In general, the historical emphasis has switched from the structure and content of Enlightenment thought, the sort of Enlightenment history written by Ernst Cassirer between the 1920s and 1940s, to the study of practices. Ideas have become embodied. The American historian Robert Darnton, for example, began innovative work in the 1960s on the multi-volume *Encyclopédie* edited by the *philosophes* Denis Diderot and Jean le Rond d'Alembert, often seen as an 'engine of Enlightenment'. Darnton treated it not as a collection of Enlightenment ideas, but as a physical entity and business venture, looking at prices and distribution, production and labour relations. In his *The Business of Enlightenment* (1979), he looked at who bought the *Encyclopédie*, and where they lived, and how its different editions gave different reading experiences for different social groups. The (often contradictory) ideas which it expressed were treated as secondary (Darnton 1979).

This was also the time when a movement born in academic literary criticism which came to be known as the New Historicism, began to exert its influence on the study of the Enlightenment. Scholars such as Stephen Greenblatt and Catherine Gallagher wanted to get away from the traditional study of the text strictly for itself which was then *de rigueur*. They founded a new journal, *Representations*, and put forward the idea that life itself was text. They operated in a new space half-way between history and literature. Influenced as well by anthropologists such as Clifford Geertz, historians such as Darnton came to look at the society of the past as of course patterned and structured, and yet as also fundamentally strange and different from the present. The scholar's task was to bring out that strangeness (see Hunt 1989; Veeser 1989). It was no longer to find causal relationships (Bonnell and Hunt 1999, 10).

The New Historicism substituted the study of practices for the study of ideas. The ideas of the Enlightenment were deprived of weight. Whereas Horkheimer and Adorno, for example, had seen Enlightenment ideas as directly operating on the world with real, albeit devastating, consequences, the New Historicists saw the ideas as being among the anthropological 'strangenesses' characteristic of the past, in other words as quite distanced from the present. It has not been until quite recently that historians of the Enlightenment have once more tried to attach values to Enlightenment ideas and to ask about their impact on the present.

The last historiographical moment which should be mentioned here is that of the so-called 'Radical Enlightenment', associated with the names of Jonathan Israel and Margaret Jacob. Jacob's work has long enlarged the canon of those seen as participating in the Enlightenment, to include freemasons, pantheists and alchemists, amongst others, without insisting on the possible links between

Enlightenment and liberal modernity (Jacob 1981). Jonathan Israel's monumental studies on the other hand, have taken the search for unorthodoxy even further with their thesis that it was the Jewish philosopher Baruch Spinoza (1632–1677) who was the central figure of a 'radical Enlightenment', rather than the thinkers of what he calls the 'moderate Enlightenment', such as Voltaire, Hume or Kant (Israel 2001, 2006, 2011). Spinoza rejected Descartes' mind-body split, and argued for the idea that God and Nature were one substance, a heretical, pantheistic idea which posed a threat not only to Christianity, but to the Deistic beliefs held by many in the 'moderate Enlightenment' as well.

Israel also argues, controversially, that his 'radical Enlightenment' lies at the foundation of the modern liberal secular theory of the state. He sees it as refusing religion, tradition, faith and authority (Israel 2011, 7). Israel's work attracted a storm of criticism, although universally commended for its industry, and its bringing to light of many lesser-known figures. But it was objected that a sharp division of thinkers into 'moderate' and 'radical' glossed over a great deal of complication and paradox in their actual thinking. Also that republicanism, and the undermining of faith, authority and tradition, 'deemed to fetter the human spirit', (Israel 2001, 686) were not necessarily part of the liberal tradition.

Enough has by now been said to make it clear that the historiography of the Enlightenment is one of unique and challenging density. I have taken the decision therefore to confine discussion to three 'hinge-periods': (1) Discussions on the meaning of Enlightenment during the European late eighteenth century; (2) debates over the meaning of Enlightenment forged against the background of the rise of the European dictatorships in the 1930s and 1940s; (3) attempts in the twenty-first century by thinkers such as the historians Jonathan Israel and Anthony Pagden, and the philosopher Richard Rorty, to see the Enlightenment as the origin of modern liberal cosmopolitanism, which is understood as a good thing.

The answer to a question

From 1784 onwards, a series of articles appeared in the journal the *Berlinische Monatsschrift*. They were attempts to answer a question on the meaning of Aufklärung (Enlightenment), which had been asked in a previous article by the Pastor Johann Friedrich Zöllner: 'What is Enlightenment? This question, which is almost as important as what is truth, should indeed be answered before one begins enlightening! And yet I have never found it answered!' Zöllner's contemporaries, however, soon rushed to his aid. Thinkers such as Immanuel Kant and Moses Mendelssohn wrote essays which were published in the *Monatsschrift*, and were closely followed by other thinkers such as Karl Leonhard Reinhold, and Christoph Martin Wieland. Easily the best-known of these articles is that by the great Prussian philosopher Immanuel Kant (1724–1804). It is frequently cited, but few readers get past the first pages, and their clarion call '*Sapere aude!*' (Dare to know!), which Kant proclaims as the motto of the Enlightenment:

> Enlightenment is mankind's exit from its self-incurred immaturity. Imma-
> turity is the inability to make use of one's own understanding without the
> guidance of another. Self-incurred is this inability if its cause lies not in the
> lack of understanding but rather in the lack of the resolution and the cour-
> age to use it without the guidance of another. *Sapere aude!* Have the cour-
> age to use your own understanding! Is thus the motto of Enlightenment!
>
> (Schmidt 1996, 58)

Here, Kant states that man should free himself from his self-imposed immaturity. To be enlightened was to learn to use reason without guidance from others.
However, later on in the essay, Kant's argument changes, and he sees unfettered
reason, unlimited questioning and defining, as potentially dissolving social, political, and religious order. Reason, rather than being the triumphant motto of the
Enlightenment, as it was a few pages earlier, becomes something to be carefully
controlled. Kant argues for the use of 'judges' to define the use of reason. As Kant
cautiously points out: 'The public use of man's reason must always be free, and it
alone can bring about Enlightenment among men; the private use of reason may
be quite often narrowly restricted' (quoted in Schmidt 1996, 59).

Kant defined the public sphere as a place where people (he defines his readers
as the '*Leserwelt*', the 'reading world') are free to write and speak critically; but
in what he calls the 'private sphere', people have a duty to restrain the expression
of unfettered judgement in order to uphold the ruler and the other powers that be,
and thus lessen the likelihood of chaos. As he continues, probably referring to
lower social classes:

> Now, a certain mechanism is necessary in many affairs which are run in
> the interest of the commonwealth by means of which some members of
> the commonwealth must conduct themselves passively in order that the
> government may direct them, through an artificial unanimity, to public
> ends, or at least restrain them from the destruction of these ends. Here
> one is certainly not allowed to argue: rather one must obey.
>
> (Schmidt 1996, 60)

Kant further clarifies his views on the public and the private with the example of
the clergyman (see Excerpt 15.1).

In making this radical division between public and private, implying that even
the pastor was 'immature', so long as he remained in his role as pastor, Kant also
implied that Enlightenment had no time for the whole moral person in political
life, an argument which has cast a long shadow. In this way, Kant tried to side-step
the potentially disruptive potential of the Enlightenment, while trying to retain its
critical impetus. Even the age's greatest philosopher saw the Enlightenment as
desirable but dangerous. As Kant concludes, it was only with a strong ruler such
as his own Frederick II of Prussia, who could create secure conditions for a controlled or limited Enlightenment, that it could happen at all.

**Excerpt 15.1 Immanuel Kant on private and
public spheres**

Schmidt, James, ed. (1996). *What is Enlightenment? Eighteenth-Century Answers
and Twentieth-Century Questions*. Los Angeles, Berkeley, CA; London: University
of California Press.

... a clergyman is bound to lecture to his catechism students and his congregation according to the symbol of the church which he serves; for he has been
accepted on this condition. But as a scholar he has the complete freedom, indeed it is his calling, to communicate to the public all his carefully tested and
well-intentioned thoughts on the imperfections of that symbol and his proposals
for a better arrangement of religious and ecclesiastical affairs. There is in this
nothing that could burden his conscience. For what he teaches as a consequence
of his office as an agent of his church, he presents as something about which he
does not have free reign to teach according to his own discretion, but rather is
engaged to expound according to another's precept and in another's name. He
will say: our church teaches this or that; these are the arguments that it employs.
He then draws out all the practical uses for his congregation from rules to which
he himself may not subscribe with complete conviction, but to whose exposition he can nevertheless pledge himself, since it is not entirely impossible that
truth may lie concealed within them, and, at least, in any case there is nothing in
them that is in contradiction with what is intrinsic to religion. For if he believed
he found such a contradiction in them he could not in conscience conduct his
office; he would have to resign. Thus the use that an appointed teacher makes
of his reason before his congregation is merely a private use, because it is only
a domestic assembly, no matter how large it is; and in this respect he is not and
cannot be free, as a priest, because he conforms to the orders of another. In contrast, as a scholar, who through his writings speaks to his own public, namely
the world, the clergyman enjoys, in the public use of his reason, an unrestricted
freedom to employ his own reason and to speak in his own person. For that the
guardian of the people (in spiritual matters) should be himself immature, is an
absurdity that leads to the perpetuation of absurdities. [60–1]

It is instructive to compare Kant's contribution to the debate triggered by Zöllner with that of the Jewish *Aufklärer* Moses Mendelssohn (1729–1786). While
Kant's audience was the *Leserwelt*, and he discussed the specific cases of preach-
ers and army officers as examples of those subject to authority, Mendelssohn
instead emphasised the need for another sort of Enlightenment, still concerned
with reason, but more conceived of as an education in the use of reason which
should spread through society. This was often called *Volksaufklärung*, or 'popular philosophy'. Yet Mendelssohn, like Kant, also made distinctions and divisions: 'Certain truths', he notes, 'which are useful to man, as man, can at times

be harmful to him as a citizen' (Schmidt 1996, 5). Both men, with very different agendas, yet worried together about the impact of Enlightenment on society and government. The German Enlightenment concerned itself, as did no other national brand of Enlightenment, with what was called 'popular philosophy', but was immediately faced with questions of class and power as it did so. German philosophy of the Enlightenment, often seen as characteristically abstract, in fact always walked a very fine line between the exploration of reason, and the stability of Church and state.

Thus, from the very beginning, most Enlightenment thinkers in Germany were unable to take ideas such as 'reason' and 'humanity' and run with them to their logical conclusion of equality. Reason, after all, was the common possession of all men. From the very beginning, Enlightenment thinkers thus regarded their own agenda in contradictory ways, and ran the free and true instantiation of Enlightenment, against the security of state, religion and society. 'Man's release from his self-incurred immaturity', as Kant put it, was accomplished by careful negotiation between the absolutes of the Enlightenment, and the needs of the polity. It was not a sudden loud cry in favour of 'Enlightenment'.

This was especially so in matters of religion. The fate of the Berlin *Aufklärer* Hermann Samuel Reimarus (1694–1768) is a case in point. His treatise, the *Apology for the Rational Worshippers of God*, argued that Revelation could add nothing to that which was already known by human reason. Reinterpreting the Gospels, Reimarus saw Christianity as a failed attempt to revitalise Judaism, and like many other Enlightenment thinkers such as Voltaire and d'Holbach, saw the modern Christian religion as a fraud perpetrated by the priesthood on the people, promising them eternal life in return for patient acceptance of the injustices of life on earth.

Enlightenment, religion, and the polity were thus intensely interacting. The death of Frederick II in 1786 only intensified the struggle. The old king, who had ruled since 1740, was personally an unbeliever, who had put in place many measures of religious toleration. He was succeeded by his devout nephew Frederick William II, whose most trusted advisers such as Johann Christoph Woellner, were opponents of the Enlightenment, and convinced of the importance of the Christian religion for the support of monarchy and state alike. Two heavily contested decrees issued by Woellner in 1788 have – perhaps ironically – some echoes of Kant's essay in the *Berlinische Monatsschrift*. The earlier decree, while allowing pastors to believe whatever they wanted in private, yet also forced them to adhere, under pain of dismissal, to the Bible and the orthodox prayer books, in their public preaching and prayers. The second decree set up an administration for the censorship of writings on religion.

The late-eighteenth-century struggle in Berlin over the relationship of Enlightenment and religion happened at the same time that in France the Revolution of 1789 moved into its opening phases. Humiliating failures in foreign policy, political and social tensions, the bankruptcy of the monarchy, and catastrophic harvests, all pushed France towards a situation where the old political structures could no longer function, and reform could only occur through radical change (cf. Chapter 16: French Revolution). How was the German *Aufklärung* to respond to this upheaval in an influential neighbouring state? It was not an easy matter to craft a response. As James Schmidt has written:

> The idea that there is a connection between the Enlightenment and the French Revolution is by now so familiar that it is difficult to imagine how troubling the relation must have seemed in the early 1790s. Because we tend to assume a natural affinity between the Enlightenment and liberal politics, we forget that many *Aufklärer* were not liberals, that some of the more ardent liberals were by no means well disposed towards the Enlightenment, and that it was by no means assumed that political revolution was a means for advancing the cause of enlightened political reforms.
>
> (Schmidt 1996, 12)

Kant's contemporaries wrestled with the problem. One of his followers, Johann Heinrich Tieftrunk (1759–1837), published his essay 'On the Influence of Enlightenment on Revolutions,' ('Über den Einfluss der Aufklärung auf Revolutionen'), in *Pharos fur Äonen* (1794), ten years after Kant's previously discussed essay, by which time the French Revolution had moved from an earlier stage of liberal reform, to one where political terror, the abolition of the monarchy, nationalisation of the French Catholic church, the execution of the king, and mobilisation for total war had brought about a profound upheaval in France. This upheaval was not merely political. It brought into question the entire moral and intellectual order. Many were there who blamed the Enlightenment. In his introduction, Tieftrunk restated this idea, commonplace in the 1790s and for long after, and much linked to such conservatives as the Abbé Barruel and Joseph de Maistre, Enlighteners who

> . . . have disparaged the religion of the people, and in this way have caused anarchy and a general corruption of morals. They bear the responsibility for all the maladies which provoked and which daily continue to provoke our age's spirit of rebellion. Enlightenment, it is said, is the source of revolutions.
>
> (Schmidt 1996, 217)

But Tieftrunk hits back at the accusers of Enlightenment:

> One seeks to make all the advances of human knowledge suspect, and for this reason one seeks to link the concept of Enlightenment to all kinds of hateful accessory concepts. Today heresy, freethinking, Jacobinism, and the rejection of all authority, however respectable, are called Enlightenment. Today, Enlightenment is treason. One must define the concept of

Enlightenment precisely and then the question can be posed: to what extent is it responsible for the events of our age?

(Ibid.)

But whose was the true Enlightenment? It is clear that Tieftrunk's audience was conceived of as male. Kant's own many disparaging comments about the capacity of other races and of the whole female sex for rational thought, and hence participation in an Enlightenment conceived of as rationality, are well known. This opened the way for a large question, one just coming to the fore in the 1790s in works such as those by the German Theodor von Hippel, *Über die bürgerliche Verbesserung der Weiber* (*On Improving the Status of Women*) (1792), the Frenchman Nicolas Caritat, Marquis de Condorcet, 'Lettres d'un bourgeois de New Haven' in *Oeuvres complètes de Condorcet* (1804), and squarely faced by the English author Mary Wollstonecraft, in her book *The Vindication of the Rights of Women*, published in 1792. How could Enlightenment confront its own ideas of the relationship between gender and rationality? As I have written elsewhere, Wollstonecraft took aim at Rousseau, and pointed out that his ideas of femininity designated women as inferior to and different from men, and

> did nothing more, as Voltaire had previously noted, than replicate in domestic life the political system based on privilege and arbitrary power, enjoyed by monarchs and aristocrats over their subjects, or slave-owners over their slaves, which those same thinkers were so ready to criticize in other contexts.
>
> (Outram 2018, 79)

After all 'Who made man the exclusive judge, if woman partake with him of the gift of reason?' (Wollstonecraft 1982, 87). Like Tieftrunk, Wollstonecraft wanted to defend the Enlightenment, but by making its philosophical basis more stable.

She pointed out a yet more serious, conceptual, problem. Enlightenment was based on ideals such as 'reason' and 'virtue', which were alleged to be innate in, or attainable by, all human beings. But rationality was precisely what was denied to women by writers such as Rousseau and Kant, and by many medical writers, while 'virtue' was defined for women in an exclusively sexual sense (Outram 2018, 79). As Wollstonecraft points out, such ideas can, however, only lead to a dangerous moral relativism which will also impede the progress of Enlightenment, by 'giving a sex to morals' (Wollstonecraft 1982, 121). She drew out the consequences:

> For man and woman, truth, if I understand the meaning of the word, must be the same; yet for the fanciful female character, so prettily drawn by poets and novelists, demanding the sacrifice of truth and sincerity, virtue becomes a relative idea, having no other foundation but utility, and of that utility, men pretend arbitrarily to judge, shaping it to their own convenience.
>
> (Ibid. 103)

And further:

> If women are by nature inferior to men, their virtues must be the same
> in quality if not in degree, or virtue is a relative idea . . . virtue has only
> one eternal standard. It is a farce to call any being virtuous whose virtues
> do not result from the exercise of its own reason . . . Contending for the
> rights of women, my main argument is but on this simple principle, that
> if she be not prepared by education to become the companion of man,
> she will stop the progress of knowledge and virtue: for truth must be
> common to all, or it will be inefficacious with respect to its influence on
> general practice.

> (Ibid. 86)

By pointing this out, Wollstonecraft not only pointed out clearly the moral relativism induced by Enlightenment beliefs about gender, and the difficulty Enlightenment had in finding a ground for morals: she also pointed out that contradiction lay at the heart of the structure of Enlightenment thought (see Excerpt 15.2) (cf. Chapter 3: Gender and social structures).

Excerpt 15.2 Mary Wollstonecraft, *A Vindication of the Rights of Women (1792)*

Wollstonecraft, Mary (1982). *A Vindication of the Rights of Women.* Ed. Miriam Kramnick. Harmondsworth: Penguin.

Rousseau declares that a woman should never, for a moment, feel herself independent, that she should be governed by fear to exercise her natural cunning, and made a coquettish slave in order to render her a more alluring object of desire, a sweeter companion to man whenever he chooses to relax himself. He carries the arguments, which he pretends to draw from the indications of nature, still further, and insinuates that truth and fortitude, the corner stones of all human virtue, should be cultivated with certain restrictions, because, with respect with the female character, obedience is the grand lesson which ought to be impressed with unrelenting rigour.

What nonsense! When will a great man arise with sufficient strength of mind to puff away the fumes which pride and sensuality have thus spread over the subject! If women are by nature inferior to men, their virtues must be the same in quality, if not in degree, or virtue is a relative idea; consequently, their conduct should be founded on the same principles, and have the same aim . . .

Probably the prevailing opinion, that woman was created for man, may have taken its rise from Moses' poetical story; yet, as very few, it is presumed, who have bestowed any serious thought on the subject, ever supposed that

---

Eve was, literally speaking, one of Adam's ribs, the deduction must be allowed to fall to the ground; or, only be so far admitted as it proves that man, from the remotest antiquity, found it convenient to exert his strength to subjugate his companion, and his invention to show that she ought to have her neck bent under the yoke, because the whole creation was only created for his convenience or pleasure.

   Let it not be concluded that I wish to invert the order of things; I have already granted, that, from the constitution of their bodies, men seem designed by Providence to attain a greater degree of virtue. I speak collectively of the whole sex; but I see not the shadow of a reason to conclude that their virtues should differ in respect to their nature. In fact, how can they, if virtue has only one eternal standard? I must therefore, if I reason consequentially, as strenuously maintain that they have the same simple direction, as that there is a God. [36–7]

If women were in fact not rational, Wollstonecraft argued, it would be far preferable to abandon pretence, and exclude them altogether from social life. To say that virtue for some human beings is not founded on rationality and is differently defined from that practised by other human beings, means it cannot grow from God, since God is one, eternal and rational.

> Without a universal, non-gendered standard of morals and rationality, it would not be possible to sustain the Enlightenment project of emancipation through universal value systems based on reason and virtue, if one half of the human race were held to lack a capacity for either quality. In other words, the way the Enlightenment thought about gender contradicted, undermined, and challenged its claims to legitimacy as a universally applicable project.
>
> (Outram 2018, 79–80)

This was an important, if not fundamental, perception about the Enlightenment, but one not carried forward by many (male) thinkers and writers who came after Wollstonecraft.

Enlightenment and totalitarianism

Discussions over the meaning of Enlightenment were thus forged against the background of the thorough-going, traumatic, violent turmoil of the French Revolution, turmoil which rapidly spread to the rest of Europe. Many authors of the late eighteenth-century, and of the Romanticism which succeeded it, were thoroughly hostile to the Enlightenment, in ways which were very similar to those still proposed a hundred years later. Herder (1744–1803) and Novalis (1772–1801) were only two of the leading names to hold this view (Ferrone and Roche 2002, 30 ff.). It was not surprising that the problematic relationship between Enlightenment and Revolution continued to pre-occupy the nineteenth century, and that the Revolution became its defining trauma. In the twentieth century, however, the questions evoked by the Enlightenment dramatically changed, as two World Wars and the Holocaust caused the violent death of millions, and catastrophe to a depth which preceding centuries had not been capable of imagining. It was not surprising that ideas which had previously been seen as basic to the western order began to be questioned. Why did Enlightenment based liberalism and humane ideals prove inadequate to stem the tide of Fascism? Why had ideals of a common 'humanity' collapsed?

During the Second World War itself (1939–1945), the stock answer to these questions among western liberal thinkers and historians such as the eminent American scholar A.O. Lovejoy (1873–1962), founder of the 'history of ideas' as an intellectual discipline, was to put the blame on Romanticism, especially German Romanticism, rather than the Enlightenment. In doing so, they made sharp distinctions between these two movements of thought, and did much to give them their separate identities. In holding and disseminating these ideas, Lovejoy was not unique, but was followed by most western thinkers. For our purposes, we will take as a representative the best-selling political theorist and poet, Peter Viereck (1916–2006), whose 1941 book *Metapolitics* carried German history from the Peace of Westphalia (1648), through to Hitler's utterances in *Mein Kampf*, to account for the emergence of Nazi ideology.

Lovejoy, Viereck, and many others of their followers, pointed out that the value of ongoing, unending 'struggle' as a creative exercise (*'streben'* in German) was central both to Romanticism and to Fascism, and that much of the conceptual underpinning of the violent nationalism which accompanied Fascism particularly in Germany and eastern Europe also came from Romanticism. Lovejoy wrote in 1941, that in this perpetual struggle:

> The nation or state itself takes on the role of the insatiable Romantic hero . . . It must ever strive for expansion, external power, and yet more power, not as a regrettably necessary means to some final rationally satisfying goal, but because continuous self-assertion, transcending of boundaries, triumph over opposition, is its vocation . . . So applied, it eventually destroyed, in many minds, the conception of a universal standard of human conduct, and the sense of a common human destiny . . . It seemed to lend a new philosophic sanction to that unreflective or animal nationalism which had long been a potent factor in European politics, but which in the *Aufklärung* had appeared to be on the wane . . . Finally, when combined with that 'permanent affective element of human nature', the collective, mutually re-enforcing amour-propre of the group, it was easily transformed into a conviction of the superiority of what is distinctive

of its 'blood', its *Volksgeist*, traditions, mores and institutions – and of its right to dominate all lesser breeds.

(Lovejoy 1941, 277)

The 'collective', Lovejoy continues, easily underpins

a national state whose members are but instruments to its own vaster ends; in which, therefore, no internal oppositions or disagreements in individual opinion can be permitted; which, however, is itself dedicated to a perpetual struggle for power and self-enlargement, with no fixed goal or terminus, and is animated by an intense and obsessing sense of the differentness of its own folk, of their duty of keeping different and uncorrupted by any alien elements, and by a conviction of the immea-surable value of their supposedly unique characteristics and cultures . . . there is a certain specific historical connection between the intellectual revolution of the Romantic period and the tragic spectacle of Europe in 1940.

(Lovejoy 1941, 278)

This was a line of thought converging with that, for example, of the bestselling American social commentator and poet, Peter Viereck, whose book *Metapolitics: From Wagner and the German Romantics to Hitler* (1941), also saw Romanti-cism, as well as the music of Wagner, racial 'science', and Fuhrer-worship, as at the origins of Nazi ideology. The nineteenth-century rejection of Enlightenment rationalism allowed ideas such as the *Volk*, the 'ground', and the 'blood' to bur-geon, and to be the justification for attacks on sovereign states whose populations contained significant German minorities. Reason, the watchword of the Enlight-enment, was thrown away (Viereck 1941, 1–9). As Hitler wrote in *Mein Kampf*: 'Civilisation means the application of reason to life. Goethe, Schiller, Kant, are reflections of the western mind. The patriot prefers to seek the "life forces", the irrational impulses, which seem to him more characteristic of the German mind' (Hitler 1940, 352).

Lovejoy and Viereck were not alone in their critique of Romanticism. Ernst Cassirer (1874–1946) the famous German Jewish philosopher of culture who emigrated to the USA in 1933 in order to escape the Nazi regime, described Romanticism as overly 'historical', and contrasted it with the Enlightenment: 'The Romantics love the past for the past's own sake . . . everything becomes understandable, justifiable, legitimated, as soon as we can trace it back to its ori-gin', whereas the Enlightenment, according to Cassirer, looks back to the past to prepare for a better future (Cassirer 1946, 180). The study of history is thus not an end in itself (see Appendix 15.1).

Cassirer castigated the work of the German philosopher Martin Heidegger, who carried on a successful academic career in the Nazi era, for weakening the power of rationality in the face of that of myth: 'We should carefully study the origin, the structure, the methods, and the technique of the political myths. We should see the adversary face to face in order to know how to combat him' (Cassirer 1946, 291).

In his earlier book, *Die Philosophie der Aufklärung*, completed in October 1932, a few months before Hitler's seizure of power, Cassirer had seen Enlightenment not just as a prop to rationality, but also as a challenge, and a source of courage. He tried to rescue it from the attacks of the romantics, and saw it as encapsulating 'reason and knowledge'. *Sapere aude!* should become our watchword too.

> . . . [wir müssen] wieder den Mut finden, uns mit ihr zu messen und uns innerlich mit ihr auseinanderzusetzen. Das Jahrhundert, das in Vernunft und Wissenschaft '[d]es Menschen allerhöchste Kraft' gesehen und verehrt hat, kann und darf auch für uns nicht schlechthin vergangen und verloren sein; wir müssen einen Weg finden, es nicht nur in seiner eigenen Gestalt zu sehen, sondern auch die ursprünglichen Kräfte wieder frei zu Machen, die diese Gestalt hervorgebracht und gebildet haben.
>
> (Cassirer 1932, 206)

> . . . we must again find the courage and daring to measure ourselves against the Enlightenment, and inwardly struggle with it. Our century, which has honored reason and science as the 'highest strength of man', cannot and must not lose its way; we must find a way not only to look it in the face, but to set free its original forces.

Cassirer died in 1946 in American exile. Within just a few years of the end of the war in Europe in 1945, however, the German philosophers Theodor Adorno (1903–69) and Max Horkheimer (1895–1973), also exiled in the US, created a completely new and influential interpretation of the Enlightenment. Reversing Lovejoy's and Cassirer's point of view, they argued that it was Enlightenment, rather than Romanticism, which was the direct forerunner of Fascism. They did so, like Cassirer, by thinking about myth, but did so in completely different ways.

Max Horkheimer and Theodor Adorno had both been members of the so-called Frankfurt School of Critical Theory, where enquiry into Fascism and other different forms of domination was fearlessly carried on until the School was forced to close, and many of its members, including Horkheimer and Adorno, fled as refugees to the USA. In 1947, just after the close of the Second World War, with Europe in ruins, Japan reeling from the first uses of the atomic bomb, and the scale of the Holocaust becoming clear, they published with the Amsterdam émigré press, Quérido, their joint work, *Dialectic of the Enlightenment* (*Dialektik der Aufklärung*) (the dialectic mentioned being the intertwining of myth and enlightenment). This work had previously been privately circulated in 1944 in a slightly different form, under the title *Philosophical Fragments* (*Philosophische Fragmente*). Their question, like Cassirer's, was 'why mankind, instead of entering into a truly human condition, is sinking into a new kind of barbarism'. This happened, according to the *Dialektik*, because of the nature of Enlightenment. Enlightenment values were always peculiarly liable to tip over into their opposite (freedom into slavery, for example):

> The Enlightenment had always aimed at liberating men from fear and establishing their sovereignty. Yet the fully enlightened earth radiates disaster triumphant. The program of the Enlightenment was the disen-chantment of the world: the dissociation of myths and the substitution of knowledge for fancy.
>
> (Horkheimer and Adorno 1987, 25; Adorno 1974, 140)

Yet Enlightenment ideas often themselves became myths, such as the very idea of 'liberty', which became the foundation myth of the new United States.

As James Schmidt has written:

> An explanation for this fatal trajectory of Enlightenment could be found in the intertwining of myth and Enlightenment that lay at the heart of what Horkheimer and Adorno came to call the 'dialectic of Enlighten-ment'... The goal of enlightenment, as they understood it, was to 'dispel myths and to overthrow fantasy with knowledge'. Yet, 'the myths which fell victim to enlightenment were themselves its products'. Enlighten-ment's attack on mythology presses forward until even its own norma-tive commitments are themselves denounced as mythical. Reason is now reduced to a strategy of self-preservation which, in the end, 'boils down to an obstinate compliance as such', that is 'indifferent to any political or religious content'. All thought that does anything other than make its peace with existing powers stands condemned as 'poetry' or empty 'metaphysics'.
>
> (Schmidt 2004, 159; original quotations from;
> Horkheimer and Adorno 1987, 21, 25, 30). (See Appendix 15.2)

Enlightenment is thus totalitarian in the sense that it abandons the quest for meaning and simply attempts to exert power over nature and the world, often through mathematical and technical means. Enlightenment rationality, Hork-heimer and Adorno argued, calibrates ends to means, and expects objectively cor-rect solutions to problems.

Yet human beings are not themselves very rational. Rationality might be a goal, but human beings find difficulty in arriving there, particularly when, as was the habit of the Enlightenment, they deprive themselves of non-rational ways to understanding, such as mythology and revelation, by calling them non-rational and hence 'un-enlightened'. The only way remaining to resolve human differ-ences in an enlightened world is through the use of force and terror. Horkheimer and Adorno thus argued that the Enlightenment had left no legacy which could resist the technologically assured man-made mass death of the Holocaust. In doing so, they were echoing arguments made by Hegel, a hundred years before. As Ferrone has pointed out, after Hegel abandoned his early support of the French Revolution: 'tragic events made it clear that the thoughts produced by the culture of the Enlightenment, in their abstract quality and claim to truth, were destined to become increasingly fantastical and polemical towards all that exists . . . the Enlightenment's dialectical movement through history ultimately resulted in the tragedy of the Reign of Terror' (Ferrone 2015, 18–19).

James Schmidt argues that Horkheimer did not need Adorno's aid to make this argument:

> The transformation of reason into a mechanism for self-preservation that paradoxically requires individuals to sacrifice themselves to the demands of the collectivity had been a persistent theme in Hork-heimer's work since the end of the 1930s. Taken by itself, this argument would have resulted in a book that repeated the argument that defenders of traditional values had long since marshalled against the Enlightenment: The grand project of freeing mankind from illusion, ultimately culminates in nihilism. What sets *Dialectic of Enlightenment* apart from arguments such as these was the other half of the chiasmus around which the plot of the book unfolded: 'Mythology is already Enlightenment'.
>
> (Schmidt 2004, 159)

Yet in spite of the contemporary appositeness of the publication of the *Dialectic of Enlightenment*, many of the problems it dealt with had been brewing since the 1930s. Many have seen the *Dialectic of Enlightenment* as an attempt by Horkheimer and Adorno to renew and revive critical theory in the light of charges that it had failed. Their concerns with human domination were constant, and in this way their interpretation of Enlightenment essentially leads on from the scorn expressed by Herder, who as we have seen, had also viewed Enlightenment as a hypocritical system of power relations. But National Socialism, Stalinism, state capitalism, mass culture, were essentially new forms of domination which had not been explained by 'traditional' critical theory. As Horkheimer's student, the well-known German philosopher Jürgen Habermas, later wrote: 'Critical theory was developed in Horkheimer's circle to think through political disappointments at the absence of revolution in the West, the development of Stalinism in Soviet Russia, and the victory of Fascism in Germany. It was supposed to explain mistaken Marxist prognoses, but without breaking Marxist intentions' (Habermas 1987, 116).

It is in this longer-term context that we can evaluate this work, as a way of thinking quite different from the literary and historical roots of Lovejoy and Viereck. And far from seeing Enlightenment as a place of daring, dynamic challenge and creativity, as had Ernst Cassirer, they saw it as an aspect of domination which had made possible the rise of National Socialism.

The contemporary liberal case

This struggle over the Enlightenment in the era of Stalinism and National Socialism, deeply torn as it was between positive and negative assessments, has been replaced by attempts to interpret Enlightenment in the light of a contemporary globalised world, and as standing at the origins of modern liberal democracy. To the fore here has been the 2013 book by the historian Anthony Pagden, already well known for his studies of the early modern Iberian world, *The Enlightenment and Why it Still Matters* (2013). Pagden begins with a disclaimer. The book is not supposed to be 'a political tract, nor a moral homily. It is a work of history . . . But all history, if it is to be anything more than mere archaeology, must be a reflection of what the present owes to the past' (Pagden 2013, xiv). However, Pagden's book can be easily placed in the very long line of commentators who have in fact inserted their own political views and aspirations into accounts of thought in the eighteenth-century. Pagden sees the Enlightenment as creating the new value of 'sympathy', which would, with relevance for a global age, link humans together by 'a passion which offered a minimum psychological principle on which to base a claim to human sociability, both within individual communities, and, what is still more significant for my purposes, across them' (Pagden 2013, 74). (See Appendix 15.3.)

Sympathy not only brings people together, but does so – and this is an important part of Pagden's argument – without religious belief:

> Sympathy gives to humankind an identity independent of God, a reason for recognizing all peoples as of equal worth, and of embracing some kind of common good, without endowing them with immortal souls or thinking of them as pale, if identical images of the divine.
>
> (Pagden 2013, 77–8)

These ideas are represented in Pagden's account by David Hume's 1748 *Enquiry Concerning Human Understanding* and his 1751 *Enquiry Concerning the Principles of Morals*, which also asserted the unity of humanity. Hume, Pagden asserts, citing his influence on Kant and Bentham, 'remains the single most influential proponent of a secular ethics based upon a "science of man" the Enlightenment ever produced' (Pagden 2013, 128).

Pagden reinforces his case by rightly rejecting the old-fashioned Romantic idea of an a-historical Enlightenment. He interprets, however, Enlightenment history-writing as revealing that the final destiny of the species must be the

> creation of a universal cosmopolitan civilization. The same goes for writings on trade and commerce, seen as linking people in reciprocal bonds. . . . It was the practical manifestation of the "sympathy" which had replaced innate sense, ideas and judgement as the one defining feature of all that it was to be human.
>
> (Pagden 2013, 151, 219–20)

Yet this civilisation needs a political form. The eighteenth century also produced powerful, durable concepts like 'patria' and 'nation', which stood at right angles to the world of cosmopolitan sympathy. As Kant put it, they inhibit 'the dissemination of general good will' (Pagden 2013, 249, 268). A world of peacefully competitive commercial states was no answer to the problem of war in a notoriously belligerent century, which was to be followed soon enough by the wars of the French Revolution and Napoleon. Some of the eighteenth-century wars were about empire and trade (ironically, Hume's 'le doux commerce'), some about dynastic or territorial objectives. It is no wonder, that in response to the horrors of war, the eighteenth century produced numerous schemes for perpetual peace, Kant's among them. None of them were adopted, though some, especially Kant's, have been seen as forerunners to the international bodies of our own day. If the Enlightenment produced a powerful drive towards the cosmopolitan unity of mankind, it certainly did not solve the problems of its instantiation.

Pagden argues that we need a state along Kant's lines, 'a truly modern state', looking like a 'modern liberal democracy' (Pagden 2013, 301–2). Kant of course had no thought of the political representation for women fundamental to the modern state. This was not only consonant with the opinions of many though not all thinkers of his age, but with his personal and often expressed views on female inferiority. Women, as irrational creatures, are not truly part of the Kantian Enlightenment defined as rationality. Pagden fails to notice this until page 309. As well, no European state until the twentieth century seriously considered including the poor of either gender as political subjects. So the sense in which Kant can actually stand as a model for the twenty-first century 'modern liberal democracy', is unclear. Pagden pays a price for his use of Kant as a model. And as we have seen, Kant's own essay on the meaning of Enlightenment stresses the fundamental conflict between unlimited thought, and political stability.

Another major problem faced by Pagden is preserving the Enlightenment from the contamination of the French Revolution: 'Any direct causal link between the Enlightenment, the Revolution, and the Revolutionary Wars which followed, although apparently obvious, is, however, a spurious one' (Pagden 2013, 328). This makes about as much sense as the argument that National Socialism had nothing to do with the Weimar Republic. It is an attempt by Pagden to get away from the nineteenth-century conservative idea that Enlightenment and Revolution were in fact tightly and destructively knit together. It is also an attempt to take the reader's attention away from the importance of the political ideas instantiated by the Revolution itself, whose 'Declaration of the Rights of Man and of the Citizen' (1789) was as least as foundational to modern liberal democracy as were the ideas of the Enlightenment. Pagden does not wish his Enlightenment to be touched by the violence and traumatic upheaval of the Revolution, which would diminish it as a model for a new cosmopolitan world order founded on 'sympathy'. Nor therefore does he want to consider the relationship between violence and the manufacture of ideology.

Pagden's often violently anti-religious positions also lead him to dubious places. He confronts for example the position of Alistair McIntyre, whose *After Virtue* (1981) argued that the Enlightenment offered no basis for the grounding of morals. This is clearly unacceptable to Pagden, since it involves seeing serious consequences for his Enlightenment's (alleged) lack of transcendence. Pagden's argument against McIntyre, which he links with an argument against communitarianism and post-modernism, is also based on personal, autobiographical grounds, on his own movements in and out of community in the course of the pursuit of his career success in professional life, making him in McIntyre's eyes, he suspects, 'a rootless and thus necessarily amoral, cosmopolitan' (Pagden 2013, 338–9). Nor does Pagden spare us his personal views on modern religion, which he calls (lumping together a universe of diverse religious phenomena) 'chaotic, chiliastic, intuitive, pathological, and for the most part utterly devoid of any theological content'. Religion, he continues, 'is most prevalent in those areas which are poor, and whose populations are undereducated'. This spiritual snobbery is no more convincing than is his explanation for the 'historical failure of Christianity to continue to provide the kind of intellectual, and consequently moral, certainty which it had once done' (Pagden 2013, 342, 343). Here Pagden's anti-Christian attitudes lead him away from the historical record, as the period of the Enlightenment which he portrays as the forerunner of unreligious liberal democracy, was in fact a period of massive religious revival (Methodism, Moravianism, Hassidism, the Great Awakening, and Pietism may be mentioned).

Pagden's book, in its historical aspects, is also prisoner to an earlier historiography of the Enlightenment, such as Peter Gay's 1980s view of it as 'modern paganism'; he discusses mainly French thinkers (no women), and says little about the social and literary practices which sustained it. Robert Darnton, for example, is absent from the bibliography. Many of Pagden's ideas have much in common with those of the American philosopher Richard Rorty (1931–2007), whom he quotes with approval (Pagden 2013, 344). To borrow some words from Rorty, modern secular liberals should believe, as Pagden does, that 'Getting rid of our sense of being responsible to something other than, and larger than, our fellow human beings is a good idea' (Rorty 2002, 20). Pagden's concern with making community in this world of contingency stems from Rorty's contention that 'Truth cannot be out there – cannot exist independently of the human mind – because sentences cannot so exist, or be out there. The world is out there, but descriptions of the world are not' (Rorty 1989, 5). These beliefs about the world have moral consequences. Rorty writes 'Nothing counts as justification unless by referring to what we already accept, and . . . there is no way to get outside our beliefs and our language . . .' (Rorty 1979, 78). In other words, we cannot escape our community, even if that community becomes immoral.

It is unclear how Rorty's philosophy of secular humanism, his 'New Pragmatism', can work as a bulwark against a repeat of dictatorship and Holocaust, which historically have overwhelmed the twentieth-century versions of – exactly – European liberal democracies. Nor is this a question, present since the 1930s, that Rorty has even recognised as existing for his project. Secular liberal democracy seems singularly unsuited to this role of resistance. Whereas Cassirer saw Enlightenment as a source of *Mut* (courage), thereby giving at least a possible balance with the Nazi emphasis on Struggle and the Will, Rorty with his New Pragmatism wants to carry forward the Enlightenment's 'project' of demystifying human life. He believed that by ridding humanity of the constricting metaphors of past traditions rooted in religion, superstition and mystification, control and subjugation, would be replaced by relationships based on freedom. He saw openmindedness as practised in Enlightenment-style tolerant conversation rather than fierce philosophical debates. But this doesn't give us the answers to questions, born in conflict and opposition, of why dictators should be opposed, or by what means. Rorty abandons the search for universal truth, in favour of what works, of the contingent. This doesn't answer these questions, or tell us how we get to a secular liberal society which may be defended. Moral identity for him is historically contingent. As some critics have noted, this all adds up to an impossible humanism, impossible because lacking any noncontingent idea of human nature. It also turns aside from a central Enlightenment tenet, that of the belief in reason. Rorty believes that 'There is no special faculty called reason allowing us to better uncover and represent the truth' (Rorty 1989, 5).

Many of the same criticisms of Rorty's ideas, may also be made of Pagden's. If we see Enlightenment as the origin of a western liberal tradition, do these arguments entrench us in a Western status quo, rather than giving us access to the (allegedly) cosmopolitan world described by Pagden? Is a commitment to liberalism itself dogmatic, a form of the cultural 'guardianship' mentioned by Kant's essay? Do both Rorty and Pagden leave us with mere caricatures of religion, for which they (inaccurately) blame the Enlightenment? Is interpreting the Enlightenment merely a project of self-creation for both Pagden and Rorty? In the end Rorty leaves us with 'hope', which allegedly survives even when Enlightenment rationalism does not. With hope we are left to extend Enlightenment into the present and future, if that is what we want to do.

Conclusion

We might want to say, with Hegel, 'when all prejudice and superstition has been banished, the question arises: Now what? What is the truth which the Enlightenment has disseminated in place of these prejudices and superstitions?' (Hegel 1967, 576). This is a question which could be asked of both Pagden and Rorty. Ultimately, the historiography of the Enlightenment is a testament to its enduring capacity to generate fundamental questions about power and historical practice, as well as at the same time, its inability to generate first-order answers to these problems.

Interpretation of the Enlightenment, both positive and negative, has been going on since the Enlightenment itself. At no time has it been divorced from questions of power, except possibly at the point at which the social practices, rather

227

than the thought, of Enlightenment began to attract the attention of historians a few decades ago. It has been uniquely often chosen by commentators such as Cassirer, Horkheimer, and Adorno, and, on another level, Pagden and Rorty, to bear the burden of prophetic commentary. The reasons for this have already been discussed. Crises of modernity have crowded in since the end of the eighteenth century. They have broken down the barriers between Is and Ought that have pre-occupied professional history since its inception in the eighteenth century. In turn, professional history has been re-shaped, and historians have become more responsive to their new position as public prophet. As the French philosopher Michel Foucault remarked, Kant's essay 'An answer to the question, What is Enlightenment?', marked 'the discrete entry into the history of thought of a question that modern philosophy has not been capable of answering, but that it has never managed to get rid of either'. It was, he added, a question which 'sought a difference: what difference does today introduce with respect to yesterday?' (Quoted in Pagden 2013, 9).

Bibliography

(A) Primary sources

Schmidt, James, ed. (1996). *What Is Enlightenment? Eighteenth-Century Answers and Twentieth-Century Questions*. Los Angeles, Berkeley, CA and London: University of California Press.

Wollstonecraft, Mary (1982). *A Vindication of the Rights of Women*. Ed. Miriam Kramnick. Harmondsworth: Penguin.

(B) Secondary literature

Adorno, Theodor W. (1974). *Minima Moralia*. Trans. E.F.N. Jephcott. London: New Left Books.

Barnard, F.M., ed. (1969). *J.G. Herder on Social and Political Culture*. Cambridge: Cambridge University Press.

Bonnell, Victoria E. and Hunt, Lynn, eds. (1999) *Beyond the Cultural Turn: New Directions in the Study of Society and Culture*. Berkeley, Los Angeles, CA and London: University of California Press.

Cassirer, Ernst (1946). *The Myth of the State*. New Haven, CT: Yale University Press.

—— (1932). *Die Philosophie der Aufklärung*. Mohr: Tuebingen.

Darnton, Robert (1979). *The Business of Enlightenment: A Publishing History of the 'Encyclopédie', 1775–1800*. Cambridge, MA: Belknap Press of Harvard University Press.

Edelstein, Dan (2010). *The Enlightenment: A Genealogy*. Chicago, IL and London: Chicago University Press.

Ferrone, Vincenzo (2015). *The Enlightenment: History of an Idea*. Princeton, NJ: Princeton University Press.

Ferrone, Vincenzo, and Roche, Daniel (2002). *L'Illuminismo nella cultura contemporanea: Storia e storiographia*. Rome-Bari: Laterza.

Habermas, Jürgen (1987). The Entwinement of Myth and Enlightenment: Horkheimer and Adorno. In: Jürgen Habermas, *The Philosophical Discourse of Modernity: Twelve Lectures*. Trans. Frederick Lawrence. Cambridge, MA: MIT Press, 106–30.

Hegel, Georg, and Friedrich, Wilhelm (1967). *The Phenomenology of Mind*. Trans. J.B. Baillie. New York and Evanston: Northwestern University Press.

Horkheimer, Max, and Adorno, Theodor (1987). Dialektik der Aufklärung. In: Max Horkheimer, *Gesammelte Schriften*, vol. 5. Frankfurt: Fischer.

Hunt, Lynn, ed. (1989). *The New Cultural History*. Berkeley, Los Angeles, CA and London: University of California Press.

Israel, Jonathan (2001). *Radical Enlightenment: Philosophy and the Making of Modernity, 1650–1750*. Oxford: Oxford University Press.

——— (2006). *Enlightenment Contested: Philosophy, Modernity and the Emancipation of Man, 1670–1752*. Oxford: Oxford University Press.

——— (2011). *A Revolution of the Mind: Radical Enlightenment and the Intellectual Origins of Modern Democracy*. Princeton, NJ: Princeton University Press.

Jacob, Margaret C. (1981). *The Radical Enlightenment: Pantheists, Freemasons and Republicans*. London: Taylor & Francis.

Lovejoy, Arthur O. (1941). The Meaning of Romanticism for the Historian of Ideas. *Journal of the History of Ideas* (June), 2 (3), 257–78.

Outram, Dorinda (2018). *The Enlightenment*. Cambridge: Cambridge University Press.

Pagden, Anthony (2013). *The Enlightenment: And Why It Still Matters*. Oxford: Oxford University Press.

Rorty, Richard (2002). The Continuity between the Enlightenment and 'Postmodernism'. In: Keith Michael Baker and Peter Hans Reill, eds., *What's Left of Enlightenment?* Stanford, CA: Stanford University Press, 19–36.

Rorty, Richard (1989). *Contingency, Irony and Solidarity*. Cambridge: Cambridge University Press.

——— (1979). *Philosophy and the Mirror of Nature*. Princeton, NJ: Princeton University Press.

Schmidt, James (2004). Mephistopheles in Hollywood: Adorno, Mann and Schoenberg. In: Tom Huhn, ed., *The Cambridge Companion to Adorno*. Cambridge: Cambridge University Press, 148–80.

Schmidt, James, ed. (1996). *What Is Enlightenment? Eighteenth-Century Answers and Twentieth-Century Questions*. Los Angeles, Berkeley, CA and London: University of California Press.

Viereck, Peter (1941). *Metapolitics: From Wagner and the German Romantics to Hitler*. New York: A.A. Knopf.

Veeser, H. Aram (1989). *The New Historicism*. New York and London: Routledge.

INDEX

For Product Safety Concerns and Information please contact our EU
representative GPSR@taylorandfrancis.com
Taylor & Francis Verlag GmbH, Kaufingerstraße 24, 80331 München, Germany

www.ingramcontent.com/pod-product-compliance
Lightning Source LLC
Chambersburg PA
CBHW060253220326
41598CB00027B/4078

* 9 7 8 1 0 3 2 0 6 4 5 4 3 *